Advanced Physical Chemistry Practical Guide

Authored by

Charu Arora

Department of Chemistry
Guru Ghasidas University
India

&

Sumantra Bhattacharya

National Institute of Technology Sikkim
India

Advanced Physical Chemistry Practical Guide

Authors: Charu Arora and Sumantra Bhattacharya

ISBN (Online): 978-1-68108-910-2

ISBN (Print): 978-1-68108-911-9

ISBN (Paperback): 978-1-68108-912-6

First published in 2022.

Bentham Science Publishers Ltd.
Executive Suite Y - 2
PO Box 7917, Saif Zone
Sharjah, U.A.E.
Email: subscriptions@benthamscience.net

BENTHAM SCIENCE

CONTENTS

FOREWORD

It is my pleasure to pen down 'foreward' of the book "The Advance Physical Chemistry Practical Guide" by learned authors – Dr. Charu Arora and Dr. Sumantra Bhattacharya. By going through the contents of the book, I can envisage that this book would be highly beneficial to students as well as professionals. I feel honoured and privileged to write preface for the book.

This book is meticulously developed as a laboratory manual devoted to Physical Chemistry and provides a wide spectrum of knowledge of the subject matter. It will help the readers to perform as well as comprehend various experiments pertaining to Physical Chemistry. The book is written in simple language so that students of all levels can easily understand the theoretical and exploratory aspects of every experiment. Beautifully detailed diagrams of experiments will further help the reader easily learn, visualize, and memorize the complicated apparatuses and equipments. During my 30 years of professional career as a physical chemist, I have observed that students always lack confidence in the analysis and presentation of data. Authors have given special emphasis on these issues. Another feature methodically included in this book is pre-lab preparation, which will certainly improve the performance and analytical skills of the students in the laboratory. In view of the recommendations of New Education Policy 2020, the authors have also accentuated the practicability of laboratory knowledge for the benefit of mankind. I am sure this book will help students in demonstrating their skills out of the laboratories and classrooms to villages and towns for the service of society.

I am confident that faculty members, students, and professionals would find this book a very useful reference treasure for their day to day working, be it teaching or research.

Both Dr. Arora and Dr. Bhattacharya deserve special compliments for their painstaking efforts to concentrate the voluminous matter in a well-structured piece of knowledge.

Keep it up! Good Luck…

Alok Mittal, D.Sc.
Department of Chemistry
Maulana Azad National Institute of Technology
(Institute of National Importance)
Bhopal 462 003
India

PREFACE

"*The test of all knowledge is experiment.* Experiment is the *sole judge* of scientific "truth""—R. P. Feynman.

Practical work has had a central and distinct role in chemistry. Experimentation is the essence of learning science. The purpose of doing experiments is to teach the principles of scientific inquiry, to improve understanding of theory through practical experience, to teach specific practical skills, such as measurement and observation, that may be useful in future studies, and to teach generic skills, such as teamwork and problem-solving, to motivate and engage students. It is important to have a clear idea of why we do practice science. It will be helpful to choose which experiments to do and how to integrate them with 'theory'. Physical Chemistry Practical Guide aims to facilitate experimental exercise in the physical chemistry laboratory at the PG level of a student's career. The book will be very helpful for teachers in providing practical knowledge of important aspects of Physical Chemistry experiments. The book covers a wide range of areas from basic to advanced experiments, including calibration of instruments as well as highly accurate software for computational quantum chemical calculations. This book has been divided into four sections: Part I consists of general introduction, calibration of glassware, instruments, and precautions; Part II entails those experiments that have a simple theoretical background and classical methods; Part III comprises of experiments that are associated with more advanced theory, and technique that requires a greater degree of experimental skill and use of instruments. Part IV comprises experiments related to the use of computers and that are investigative in nature. Covering all aspects of classical, advanced, and computational chemistry experiments, this book will be useful for under graduate and post-graduate students to gain confidence in their ability to perform a physical chemistry experiment and to appreciate the value of the experimental approach.

We also celebrate this opportunity for expressing bottom-hearted gratitude towards the people who supported us at all stages of our work. The authors acknowledge their parents, spouse, family members, friends, and colleagues for their continuous support and encouragement above all our students. The basis of the book is to overcome difficulties that arise while performing physical chemistry practicals, and this we learned from our experience during our studenthood as well as interacting with the students during practical classes. We would like to acknowledge Mr. Sanju Soni, Ph.D. student of Department of Chemistry, Guru Ghasidas University, Bilaspur, for his dedicated efforts in preparing figures/improving the quality of figures in the book chapters without which this book would not have become a reality. We are also thankful to Dr. Amlan Das, NIT Sikkim (National Institute of Biomedical Genomics., Kàlyani) for drafting experiment on the denaturation of Bovine Serum Albumin (Protein) and Mr. Happy Mondal from NIT Sikkim for designing experiments on Determination of Hall coefficient of a semiconductor and determination of paramagnetic susceptibility of a given paramagnetic material. We would like to express our gratitude to Bentham Science for publishing the book.

CONSENT FOR PUBLICATION

None.

CONFLICT OF INTEREST

The authors declare no conflict of interest.

ACKNOWLEDGEMENTS

The authors would like to acknowledge the Vice Chancellor, Guru Ghasidas University, Bilaspur, and Director, NIT, Sikkim, for providing necessary facilities. The authors acknowledge their parents, spouse, family members, friends, and colleagues for their continuous support.

Charu Arora
Department of Chemistry
Guru Ghasidas University
India

&

Sumantra Bhattacharya
National Institute of Technology Sikkim
India

CHAPTER 1

General Introduction

UNITS AND SIGNIFICANT DIGIT

List of Units Useful for This Book

The concentration of a solution is a measure of the amount of solute which has been dissolved in a given amount of solvent or solution. A concentrated solution has a relatively large amount of dissolved solute. A dilute solution has a relatively small amount of dissolved solute.

Molarity (M) is one of the most common methods to express concentration. Molarity (M) indicates the number of moles of solute per liter of solution (moles/Liter) and is one of the most common units used to measure the concentration of a solution. Molarity can be used to calculate the volume of solvent or the amount of solute.

Molality (m) is another way to represent the concentration of a solution and is defined as the number of moles of solute per kilogram of solvent. The SI unit for molality is mol/kg. For example, a solution with a molality of 3 mol/kg is often described as "3 molal" or "3 m." However, following the SI system of units, mol/kg or a related SI unit is now preferred.

Equivalent weight (also known as gram equivalent) is the mass of one equivalent, which is the mass of a given substance that will combine with or displace a fixed quantity of another substance. The equivalent weight of an element is the mass that combines with or displaces 1.008 grams of hydrogen or 8.0 grams of oxygen or 35.5 grams of chlorine. All these are summarized in Table **1.1.**

Table 1.1. Units of various parameters used in the book.

Parameter	CGS Unit	SI Unit
Concentration	g/cm^3, M	Kg/m^3, mol/kg, M
Molar absorption coefficient	$M^{-1}.cm^{-1}$	$M^{-1}.m^{-1}$
Unit of rate constant (First order reaction)	s^{-1}	s^{-1}
Viscosity	Poise or $Dyncmsec^{-1}$	N-m/sec
Surface tension	$Dyncm^{-1}$	Nm^{-1}
Specific Rotation	$Deg.cm^{-2}g^{-1}$	$Deg.m^{-2}kg^{-1}$

Charu Arora and Sumantra Bhattacharya

Conductance	Ohm^{-1} or mho siemens	-
Potential difference	Volt	Volt
Universal molar gas constant	1.987 $calmol^{-1}K^{-1}$	8.314 $JK^{-1}mole^{-1}$

Significant Digit

In measurement reporting, a number with a proper decimal place is important. Before representing a value, one experimentalist should know the precision of instruments and apparatus used. To determine how many significant figures are in a number one should follow these rules:

1. Non-zero digits are always significant.

e.g., 123 (3 significant figures)

2. Any zeros between two significant digits are significant.

e.g., 11.102 (5 significant digits)

3. Zeros to the left of the first non-zero digit are not significant.

e.g., 0.05 (1 significant figure)

4. If a number ends in zero to the right of the decimal point, those zeros are significant.

For example, 2.00 (3 significant digits)

5. If a number ends in zeros to the left of the decimal point, those zeros are not essentially significant. *e.g.*, If we make a statement that the reading of a burette in some titration process is 12.10 ml. It is good to report more digits after the decimal place, but simultaneously one should keep in mind that any observation can be reported up to the least count of the apparatus. Since generally burette shows one point after the decimal place. Thus, it is not recommended to report beyond the first digit of the decimal place. On the other hand, when an experimentalist is multiplying or dividing two numbers obtained from different apparatus with different least counts, then the experimentalist should report the value having no more significant figures than the least accurate observation.

e.g., If an object has mass of 29.1143 g and a volume of 25.0 cm^3, then its density is given by 29.1143 g/25.0 cm^3 = 1.164572 g/ cm^3 =1.2 g/cm^3

Here please note that if we report only up to the first digit after the decimal, then the final value should be 1.1. Since the digit at the second decimal place is more than 5, so we have used 2 instead of 1. This method is called the round-off method. However, you are required to round off numbers only at the END of calculations; otherwise, errors may be inadvertently carried through.

CALIBRATION OF VOLUMETRIC APPARATUS

Since all the volumetric analyses are calculated based on burette reading thus, it is necessary to calibrate the value of all the apparatus, like a pipette, volumetric flask, *etc.*, in terms of burette reading. To illustrate this, let us pick up an example of calibration of the pipette (10 ml). To do so, let us fill up the burette (50 ml) with distilled water up to the mark and then elute the known volume of water and note the reading. Then pipette out water using the pipette to be calibrated. Pour the water into the burette and note the reading and difference. The difference between these two values gives the measurement of the calibration of the pipette. Similarly, other apparatus can also be calibrated.

CALIBRATION OF BALANCE

Calibration weights are specially designed cast iron or stainless-steel weights used to calibrate weighing equipment. For external calibration of a balance, the user must have a set of approved weighing scale calibration weights that should be kept in top condition. The scale calibration weights are put on the balance or scale and their mass or weight is set as the standard.

PREPARATION OF STANDARD SOLUTION

For several titrimetric processes, it is necessary to prepare standard solutions, which are of two types, *viz.* primary standard, and secondary standard. The strength of primary standard solutions is maintained for a long time. While performing an experiment, it is necessary to prepare the primary solution very accurately. To do so, one should weigh the solute very accurately. Pour the solute into a volumetric flask and add solvent up to the mark. Shake the solution very well to make a homogeneous solution. While adding solvent, one should be very careful to avoid the addition of excess water. If mistakenly added, then note down the amount of excess water added and calculate the strength accurately. In this context, it is quite difficult to make an exact solution. So, the coefficient should be calculated accurately. To illustrate this, let us prepare 100 ml 0.1 (M) oxalic acid ($H_2C_2O_4.2H_2O$) solution, *i.e.*, 1.26 g of $H_2C_2O_4.2H_2O$ is required. Someone weighed 1.35 g of $H_2C_2O_4.2H_2O$ and added 101 ml of water instead of 100 ml of water. Thus, the strength of the solution will be 0.108 (M).

STATISTICAL TREATMENT OF DATA ANALYSIS

Accuracy and Precision

When an observed value is close to the actual or acceptable value, this is known as accuracy.

Precision

Precision is how close the measured values are to each other.

Examples of Precision and Accuracy

Accuracy and precision have been represented in Fig. (**1.1**).

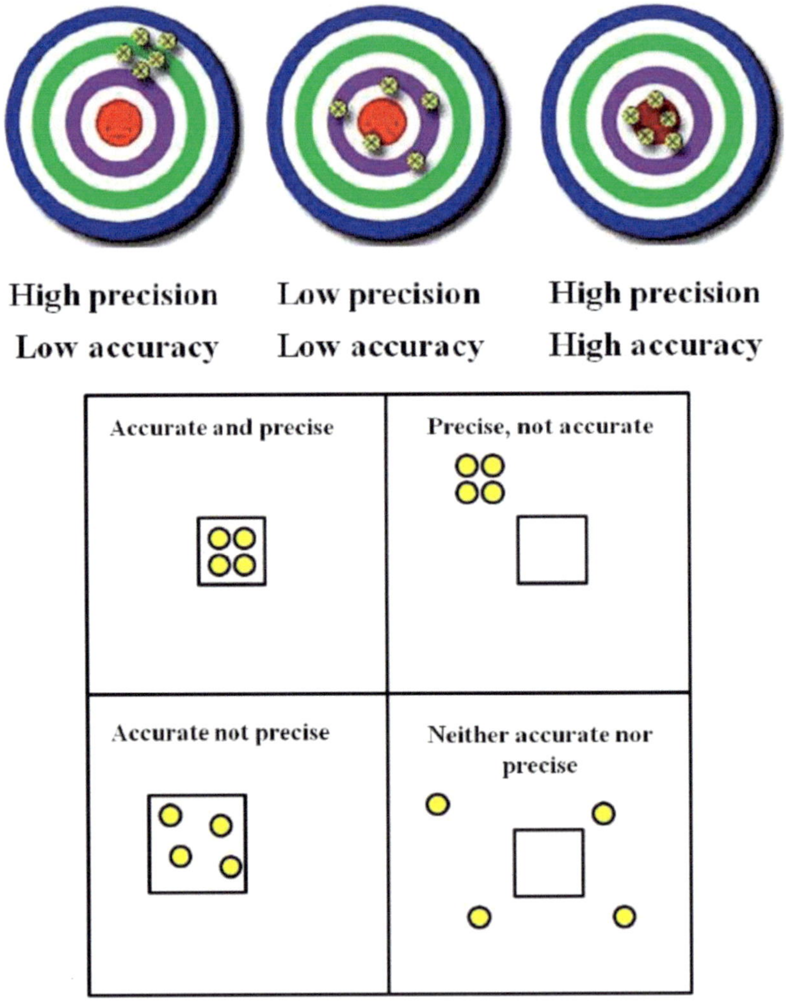

Fig. (1.1). Pictorial representation to exhibit the difference between accuracy and precision: they mean slightly different things!

ERROR

The difference between the measured value and the actual value is defined as an error. The Error can be classified as:

(1) Personal error is the one that arises because of faulty procedure adopted by the observer, *e.g.*, making an error in reading a scale.

(2) Systematic error can be caused by a defect in the measuring device. This error can be removed by proper calibration of the instrument.

(3) Random Errors are beyond the control of experimentalists, *e.g.* sudden change in experimental conditions, like fluctuation in current in conductometric or potentiometric measurement.

PRECAUTIONS IN LABORATORY

General Rules and Safety Precautions to Be Taken in a Chemistry Laboratory

It is necessary to follow some necessary precautions to avoid the chance of an accident and run the chemistry laboratory class smoothly.

1. Entry with a lab coat as well as a shoe in a chemistry laboratory is recommended.

2. To avoid exposure to hazardous chemicals, use gloves, masks, and safety glasses.

3. Please avoid using mobile phones while doing experiments.

4. Clean the apparatus; glassware should be cleaned and dried properly before the experiment. Students are advised to clean the apparatus, glassware as well slab once the experiment is over.

5. Students are advised to dispose of solid non-harmful waste in the dustbin, hazardous wastes in the proper chemical waste bin. In addition, make sure no piece of broken glassware is on the floor or slab. In case of glassware breakage, please inform the lab attendant and clean them immediately.

6. Switch off electrical or electronic apparatus after their use.

7. Students are advised to protect their eyes or keep themselves at a safe distance while using hazardous chemicals, like, ammonia, hydrochloric acid, *etc.*

8. Students are advised to update lab notebooks regularly.

9. Teachers and instructors are advised to arrange fire safety measures as well as the first aid box in the laboratory.

Precautionary Rules

1. Don't pipette out strong acids and bases by sucking; it can be dangerous, therefore, use measuring cylinders or droppers for such chemicals.

2. Please make sure that chemical bottles are properly covered, else it will lead to unnecessary loss of chemicals as well as laboratory environment can be full of harmful fumes.

3. Avoid using a contaminated spatula or any other materials.

4. In case of an accident, do not panic. Please report your teacher or lab attendant immediately and do not forget to consult a physician for treatment.

5. Avoid contact with any chemical in your body or clothes.

GENERAL INSTRUCTIONS

Safety in the Chemical Laboratory

It is required to exercise due care and caution to avoid accidents in the laboratory. Proper handling of apparatus and chemicals is the only insurance against hazards. In particular-

i) Do not throw away lighted match-sticks without extinguishing them.

ii) Do not touch any unknown chemical by hand.

iii) Ensure that gas taps are securely closed while putting off burners.

iv) In case of any exposure to corrosive chemicals such as strong acids, wash the affected part in flowing water, and contact the instructor for further help.

v) Never try to neutralize the corrosive substance chemically, as this may result in burns. In case of any exposure, move away from the source of exposure (or remove the source).

Maintenance of Laboratory Records

i) In any experiment, the results are to be faithfully recorded and analyzed. The maintenance of laboratory records is an essential part of all scientific work.

ii) The experimental observations must be taken down in a clean bound notebook and not in loose sheets and certainly not in the margins of the instruction's sheets.

iii) There should be no erasure in the observation book.

iv) If any entry needs to be corrected, it must be crossed out neatly and a new entry is made.

v) After the day's experimental work is over, get the observations initiated by the instructor. If the time is left over, do the necessary calculations and get them checked too. Write up the work in the fair laboratory record with details of the principle and procedure involved, neatly tabulate the data and calculations, and present the report on the next turn regularly.

vi) The maintenance of a day-to-day observation book, complete in all details, is an essential part of any scientific work at any level and is a habit that you must cultivate.

General Organization of Laboratory

In the few weeks, you will do a set of common experiments involving titrimetry and learn (if you are not already familiar with them) the usage of the electronic balance, the techniques of titrimetry, and general skills of handling glassware. Thereafter, you will do other experiments in rotation and for that, seating arrangement will be displayed. Some experiments or chemicals may be issued to you from the counter for every experiment; you must return them faithfully at the end of your experiment so that the next student can use them. Any breakage must be reported to the laboratory assistant, and a replacement should be obtained.

Read the instructions carefully and apply your mind while doing the experiments. Nothing is gained by merely 'following the experiment'. The purpose of the laboratory work is to understand better the chemical principle involved.

Volumetric Analysis

In volumetric analysis, we compare the strength of two solutions by finding the volume of one that exactly reacts with a known volume of the other. To determine the strength of one of the solutions, the other solution of known strength is required. We call such a solution a 'Standard Solution'. The act of determining the strength of the solution is called standardization. To standardize a solution volumetrically, a solution whose strength is known, independent of titration, is required. A substance whose standard solution may be prepared without standardization by titrimetry is called primary standard. For usefulness, a primary standard must not change its strength under storage. Sodium hydroxide is a common chemical that cannot be used as a primary standard because

(i) Both the solid and its solution absorb CO_2 from the atmosphere.

(ii) The solution tends to dissolve silica from glassware used for storage.

Oxalic acid, on the other hand, can be used as a primary standard. Commonly, primary standard solutions are prepared by dissolving known weights of pure solids (Analytical grade) reagent and making the solution up to a known value.

Preparation of Standard Solution

The proper handling of the chemical/electronic balance and the volumetric flask must be learned. Let us say that we want to weigh out a known quantity of a solid 'A' and dissolve it in water and make up 100 ml of solution. Collect the weighing bottle from the laboratory assistant. Take 100 ml volumetric flask and funnel supplied to you, place the funnel in the mouth of the flask and take the whole assembly to the weighing room. Place approximately the required quantity of a substance 'A' in the weighing bottle, and weigh it on the electronic balance. Important: Electronic balance should be handled very carefully, and no chemical should be spilled in the pan. Remove the substance from the pan and nearby surface if spilled by chance.

Weighing

Note the weight of the bottle and contents (say X g). Now take the bottle out, and tap out its contents into the funnel. Do not worry about the crystals that stick to the sides of the bottle. Weigh it again (say Y g), then X-Y gives the weight of the substance transferred. Using a wash bottle, wash down the contents of the funnel into the flask, ensuring that no splashing occurs from the funnel surface. Wash the funnel repeatedly all around by giving it a circular motion while using the wash bottle.

Caution

Take care that the contents of the flask do not come above half the bulb of the flask. Remove the funnel and swirl the flask till all the crystals are dissolved. Add distilled water till the level reaches the neck (but below the mark) and swirl to ensure mixing. Now raise the flask till the mark is at eye level, and add water in the drops till the lower meniscus reached the mark. Stopper the flask and shake thoroughly. Now you have 100 ml of a standard solution of 'A'.

Handling of Pipette and Burette

The volumetric flask is designed to contain a known volume of liquid, whereas, the pipette and burette are designed to deliver a known volume, at a temperature that is marked on the bulb of the pipette or near the top of the burette. They deliver the correct volume of liquid only when the correct technique is used. The burette should not be drained fast. All reading must be taken with a burette strictly vertical and the meniscus at eye level. For colorless solutions the lower meniscus is read and for intensely colored solutions, read the upper meniscus. Before use, the burette is cleaned and rinsed with the solution to be used. The solution is run off fast till nay air.

Handling of Density Bottle and Pyknometer

Density bottles are mainly used to determine the density of liquids of moderate viscosity. They are not volumetric instruments, however, they are calibrated 'to contain' as in the case of volumetric flasks. Pyknometer is a standard vessel for measuring and comparing the densities of liquids or solids. The volume of the Pyknometer is precisely known. Pyknometer is used to determine the density of a liquid by filling the container with the liquid and then weighing it, whereas a density bottle is any bottle of known volume that can be weighed empty and then weighed containing the liquid.

The weight of the dry and empty density bottle is determined. Fill the density bottle with liquid, avoiding bubbles. The ground neck should be covered to about 1/3. In a thermostatic bath, adjust the temperature of the bottle and contents to 20° C. Align the stopper respectively, the thermometer of the density bottle according to the marking, and insert carefully. The capillary tube fills up, and the displaced liquid comes out. Carefully dry the outer surfaces of the stopper (as well as, the side capillary) and the density bottle with tissue. Determine the weight of the filled density bottle.

FURTHER READING

[1] Christian, G. D. "Analytical Chemistry" 7th Edition; John Willey & Sons INC; USA, 2014.

[2] Jeffery, G. H.; Bassett, J.; Mendham, J.; Denney, R. C.; "Vogel's Text Book of Quantitative Chemical Analysis" 5th edition; Longman Scientific & Technical; Great Britain, 1989.

CHAPTER 2

Physical Chemistry Practical using Thermo-Chemistry

INTRODUCTORY REMARKS

Thermochemistry deals with heat changes taking place during chemical reactions as well as the heat changes associated with physical transformations. Reactions accompanying evolution or absorption of heat are known as exothermic or endothermic reactions, respectively. The heat of reaction for a chemical process depends upon the following factors:

1) Nature of reacting substances.

2) Physical state of reacting substances (*e.g.*, solid, liquid, and gaseous state are a particular allotropic form).

3) The fraction of reactive species taking part in the reaction.

Heat changes are generally expressed in calories (cal.). It is defined as the amount of heat required to raise the temperature of 1 g of water from 15° C-16° C. Another conventional unit is the joule (J) (Came by the name of James Prescott Joule). 1 cal. = 4.184 J.

A thermochemical result is expressed by writing the chemical formula of reacting substances separated by comma (,) together with the amount of heat change that has accompanied the reaction. A comma is used to separate the chemical formula of reacting species when they are combined directly, while a colon (:) when the reaction takes place, but there is no direct combination. Thus, (N, 3H) indicates that nitrogen and hydrogen react directly to form NH_3, while (NH_3: $3Cl_2$) indicates that ammonia and chlorine react to form HCl and NCl_3.

Thermochemical variables can be classified as follows:

i) Heat of formation

ii) Heat of solution

iii) Heat of dilution

Charu Arora and Sumantra Bhattacharya

iv) Heat of hydration

v) Heat of neutralization

vi) Heat of reaction

vii) Heat of combustion

The demonstration of measurement of all these is beyond the scope in this book. We will demonstrate only Heat of dilution, Heat of neutralisation and Heat of reaction.

DESCRIPTION OF CALORIMETER

A calorimeter is a device used to measure the heat flow of a chemical reaction or physical change. The process of measuring this heat is called calorimetry. A basic calorimeter consists of a metal container of water above a combustion chamber, in which a thermometer is used to measure the change in water temperature. However, there are many types of more complex calorimeters.

The basic principle is that heat released by the combustion chamber measurably increases the temperature of the water. The temperature change may then be used to calculate the enthalpy change per mole of substance A when substances A and B are reacted.

The equation used is:

$$q = C_v*(T_f - T_i)$$

Where, q, C_v, T_f, and T_i are the amount of heat in joules, calorimeter's heat capacity in joules per Kelvin (J/K), final and initial temperatures of the system.

I. TO FIND WATER EQUIVALENT OF CALORIMETER AND DETERMINATION OF HEAT OF DILUTION OF H_2SO_4

Chemicals and Apparatus

Sulphuric acid, oxalic acid, sodium hydroxide, distilled water, calorimeter.

Theory

Heat capacity or water equivalent of a calorimeter is defined as the number of calories required to heat the calorimeter by unit temperature. If M is the mass of the calorimeter and S is the specific heat then heat capacity is obtained by multiplying M by S. During the heat changes the calorimeter takes up some of the heat evolved, it should be taken into account by determining the water equivalent (w.e.) or heat capacity of the calorimeter.

In the case of glass vessels, the value of w.e. is found for such part of the vessel which is actually in contact with the reacting system. In this case, the method of obtaining w.e. by multiplying the mass and specific heat of the material of the vessel is not significant. During the experiments, equal volumes are used so that the area of the calorimeter in contact with the system remains unaltered as far as possible.

The heat of dilution is a quantity of heat evolved or absorbed when a solution containing 1 g-mol of a substance in an unknown quantity of water or other solvent is diluted by the known quantity of that solvent. Students should keep in mind that the dilution process is not a chemical reaction.

Procedure

1) Calculate the density of water at two different temperatures, sulphuric acid using the method described in Experiment 2, Chapter 4.

2) Take 25 ml of distilled water in the calorimeter and record its temperature. Take some water in a beaker and heat it to a temperature of about 30° C-35° C, higher than the room temperature.

3) Pipette out 24 ml hot water and add it to another beaker and record its temperature after every half minute for five minutes.

4) Add this hot water quickly to the water in the calorimeter.

5) Mix the content properly and record the temperature after every half minute.

6) Plot a graph of temperature *vs.* time and from it find out the temperature of hot water and that of the mixture at the time of mixing.

7) On the graph draw a vertical line for the moment of mixing (when the half volume of water has been transferred) and extrapolate the temperature-time curve

of hot water and mixture to this vertical line. The point of intersection gives the desired temperature (Fig. **2.1**).

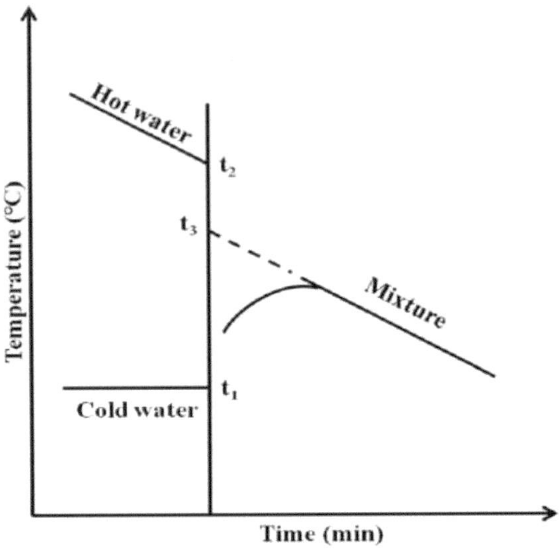

Fig. (2.1). Experimental plot Time *vs*. Temperature.

8) Now perform another experiment and take 25 ml distilled water in the calorimeter and note its temperature.

9) Pipette out known quantity (say 5 ml) of sulphuric acid of known strength. Transfer this acid into the calorimeter and mix the content of the calorimeter with the pipette itself. Record the temperature with time as done in the previous part of the experiment and plot the temperature-time curve. Note the maximum temperature attained.

Observations

Temperature = ……°C

W_1= weight of density bottle/pyknometer

W_2= weight of pyknometer + water

W_3= weight of pyknometer + liquid

Therefore, $(W_2 - W_1) = W$= weight of water at t°C

$(W_3 - W_1) = W' =$ weight of liquid at t°C

Thus, the Density of liquid/water $(d_t) = (W'/W) * D$.

I) General Observations

The volume of cold water = V_1 ml.
The initial temperature of water = t_1° C.
The volume of hot water mixed = V_2 ml.
The temperature of hot water = t_2° C.
The temperature of mixed solution = t_3° C.

II) Calculation of Density (D) of Liquid/Water

Volume of cold water taken = V_3 ml.
The initial temperature of water = t_1° C.
The volume of sulphuric acid = V_4 ml.
Highest temperature after addition = t_4° C.
Strength of sulphuric acid = x (M).

Calculations

I) Heat taken by calorimeter and water = $(W + V_1 \times d_1) \times (t_3 - t_1)$ cal.
Here d_1 is the density, W is the weight of the calorimeter.
Heat given out by hot water = $V_2 \times d_2 \times (t_2 - t_3)$
The heat taken up = heat given up

Thus, $(W + V_1 \times d_1) \times (t_3 - t_1) = V_2 \times d_2 \times (t_2 - t_3)$

$$W = \frac{V_2 \times d_2 \times (t_2 - t_3) - V_1 \times d_1 \times (t_3 - t_1)}{(t_3 - t_1)}$$

II) Heat of dilution of x (M) sulphuric acid = $(W + V_3 d_1 + V_4 d_4) \times (t_4 - t_1)$ cal.

Results

Heat capacity of the calorimeter = ... Cal.

The heat of dilution of sulphuric acid = ... Cal

Precautions: 1) The temperatures should be recorded accurately up to the least count of the thermometer.

2) The calorimeter should be completely insulated.

II. TO FIND OUT HEAT OF NEUTRALISATION OF NaOH AND HCl

Chemicals and Apparatus

Oxalic acid, HCl and NaOH, calorimeter, thermometer, stirrer and burette, pipette, measuring flask.

Theory

Heat of neutralisation is the quantity of heat evolved on neutralisation of 1 g-equivalent (g-eqv.) of acid by 1 g-eqv. of a base in dilute solution. When dilute solutions of strong acids are neutralised by strong base of about the same concentration. The heat evolved is found to be practically constant (13.7 kcal) for all strong acids and bases. Strong acids and bases in their dilute solutions are almost completely ionized and on neutralisation, they result in the formation of salt, formed by their union, so that the only change can be said to be the formation of water by reaction of H^+ and ^-OH ions, as shown below:

$$NaOH + HCl = NaCl + H_2O + 13.7 \text{ kcal}$$

$$\text{or, } Na^+ + {}^-OH + H^+ + Cl^- = Na^+ + Cl^- + H_2O + 13.7 \text{ kcal}$$

$$\text{or, } H^+ + {}^-OH = H_2O + 13.7 \text{ kcal}$$

It is obvious from the above reactions that the heat of neutralisation of a strong acid by strong base represents the heat of a combination of 1 g-eqv. H^+ ion with the same g-eqv. of ^-OH to form water.

Procedure

1) Determine the water equivalent of the calorimeter as described in the Expt. 1.

2) Prepare (M/2) oxalic acid by dissolving 15.75 g oxalic acid in 250 ml distilled water.

3) Prepare ~ 2 (M) NaOH by dissolving 85 g NaOH in 1 liter.

4) Standardise the NaOH solution against oxalic acid and prepare exact 1 (M) 250 ml NaOH.

5) Prepare nearly 2 (M) HCl by diluting 50 ml concentrated HCl to 250 ml and standardize it by titrating against standard 1 (M) NaOH.

6) Prepare exact 1 (M) 250 ml HCl by proper dilution. Recheck it by titrating against standard NaOH solution.

7) Take 100 ml 1 (M) HCl in the calorimeter and record the temperature after every half minute for five minutes. Repeat the procedure for 100 ml 1 (M) NaOH and plot the temperature-time curve for both experiments.

8) Pour NaOH quickly into the calorimeter containing HCl, taking care to avoid splashing. Mix it properly and record the temperature readings after every half minute for five minutes. After the completion of experiments add 1 drop of phenolphthalein to ascertain the complete neutralisation.

9) Plot a graph between temperature and time (Fig. **2.2**).

10) It has been observed that after mixing HCl and NaOH the temperature rises quickly in the beginning, as energy (heat) is produced when an acid reacts with a base in a neutralisation reaction. The temperature rise is irregular in the beginning and later on, it increases in a regular manner. Record the temperature of the mixture at the time of mixing.

11) Draw a vertical line at the time of mixing (when a half quantity of NaOH has been added). Extend the curves through the points when the temperature begins to fall regularly. The point of intersection will give the final temperature after mixing.

12) To obtain the actual temperature at the time of mixing plot the temperature *vs.* time for each solution. The temperature of solution before mixing can be taken as: $\left(\frac{t_1+t_2}{2}\right)$, as they are mixed in equal volumes.

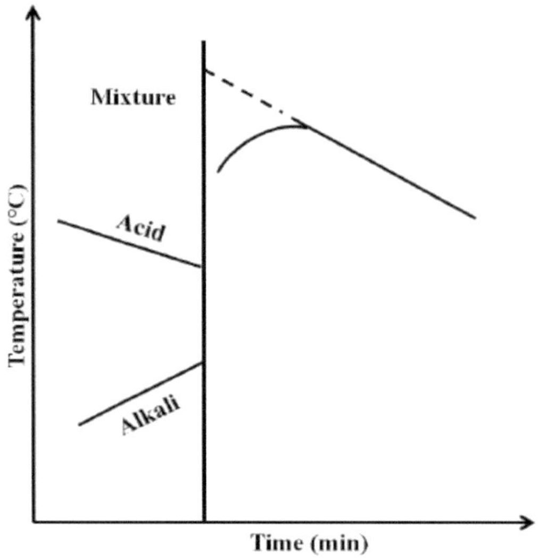

Fig. (2.2). Plot of temperature *vs.* Time.

Observations

Volume of cold water =V_1 ml

Initial temperature = t_1° C

The volume of hot water added = V_1 ml

The temperature of hot water = t_2° C

Final temperature after mixing = t_3° C

Calculation

(I) If W is water equivalent of the calorimeter, then, heat taken by calorimeter and water = $(W+ V_1) \times (t_3 - t_1)$ cal.

The heat released by hot water = $V_2 \times (t_2 - t_3)$ cal

Heat has taken up = heat given out

$$(W+ V_1) \times (t_3 - t_1) = V_2 \times (t_2 - t_3)$$

$$\text{Or, } W = \frac{V_2\times(t_2\text{-}t_3)\text{-}V_1\times(t_3\text{-}t_1)}{(t_3\text{-}t_1)}$$

Rise in temperature = $(t_5\text{-}t_4)°$ C

Heat is given out by the solution = $(V_3+V_4+W) \times (t_5\text{-}t_4)$=Q cal

Therefore, Q cal. of heat is given out by reaction of 0.1 (M) of HCl and 0.1 (M) NaOH.

So, molar heat of neutralisation = $\frac{Q\times 1\text{mole}}{0.1}$ cal = $(10\times Q)$ cal.

Result

Heat of neutralisation of HCl and NaOH = ... Cal.

III. TO FIND RELATIVE STRENGTH OF TWO ACIDS BY ESTIMATING HEAT OF NEUTRALISATION

Chemicals and Apparatus

Oxalic acid, NaOH, HCl, and acetic acid, calorimeter set up.

Theory

The strength of an acid or is the measure of its extent of ionization (degree of dissociation) at a given concentration. If the heats of neutralisation of HCl acid (here onward hydrochloric acid or HCl acid will be denoted as HCl only) and acetic acid against NaOH are x and y calories, respectively. If to a mixture containing 1 g-eqv. of each acid, 1 g-eqv. of NaOH is added, HCl and acetic acid will consume NaOH in the ratio of their ionization. Assuming n g-eqv. out of 1 g-eqv. of NaOH is consumed by HCl and remaining (1-n) by acetic acid, nx and [(1-n)×y] cal will be produced by neutralisation of HCl and acetic acid, respectively. Suppose z is experimental heat evolved by the mixture, then:

$$nx+(1\text{-}n)y = z$$

$$\text{or, } n = \left(\frac{z\text{-}y}{x\text{-}y}\right) \qquad \textbf{(2.1)}$$

The ratio: n: (1-n) gives the relative strength of HCl and acetic acid.

Procedure

1) Determine the water equivalent of the calorimeter as described in Expt. I.

2) Prepare (M/2) oxalic acid by dissolving 15.75 g oxalic acid in 250 ml distilled water (Table **2.1**).

3) Prepare ~ 2 (M) NaOH by dissolving 85 g NaOH in 1 liter.

4) Standardise the NaOH solution against oxalic acid (Table **2.2**) and prepare exact 1 (M) 250 ml NaOH.

5) Prepare nearly 2 (M) HCl by diluting 50 ml concentrated HCl to 250 ml and standardize it by titrating against standard 1 (M) NaOH (Table **2.3**).

6) Prepare exact 1 (M) 250 ml HCl by proper dilution.

7) Prepare ~2 (M) acetic acid by diluting 30 ml glacial acetic acid to 250 ml and standardize it by titrating against standard 1 (M) NaOH (Table **2.4**).

8) Prepare exact 1 (M) 250 ml acetic acid by proper dilution.

9) Take 100 ml of HCl in the calorimeter and neutralise it with the same quantity of NaOH solution. Record the heat of neutralisation for this process. This gives x.

10) Repeat Step 7 with acetic acid. This gives y.

11) Take 125 ml of each of 1 (M) acetic acid and HCl in a 250 ml volumetric flask and shake thoroughly to make a uniform mixture. This mixture contains 0.5 g-eqv. per liter of each acid.

12) Add 250 ml of 0.5 (M) NaOH to this mixture and determine the heat evolved during the reaction. On multiplying this quantity by 8 (the value of heat evolved for the solution containing 1 g-eqv. in 1 liter), we can evaluate z.

13) Calculate the ratio n:(1-n) using Eq. (2.1).

Observations

Table 2.1. Preparation of (M/2) 250 ml oxalic acid solution.

The molecular weight of oxalic acid ($C_2O_4H_2.2H_2O$=126.03)

Initial Mass (g)	Final Mass (g)	Mass of Oxalic Acid Transferred (g)	Mass of Oxalic Acid to be Taken (g)	Strength of Oxalic Acid Solution (M)
-	-	-	15.75	-

Table 2.2. Standardization of NaOH against oxalic acid.

Sl. No.	Vol. of Oxalic Acid Taken (ml)	The Volume of NaOH Consumed (ml)	Most Precise Reading (ml)
1	10	-	-
2		-	-
3		-	-

Table 2.3. Standardization of HCl acid against NaOH.

Sl. No.	Vol. of Sulphuric Acid Taken (ml)	The Volume of NaOH Consumed (ml)	Most Precise Reading (ml)
1	10	-	-
2		-	-
3		-	-

Table 2.4. Standardization of acetic acid against NaOH.

S. No.	Vol. of Sulphuric Acid Taken (ml)	Volume of NaOH Consumed (ml)	Most Precise Reading (ml)
1	10	-	-
2		-	-
3		-	-

Result

Thus, the heat of neutralisation of mixture of HCl and acetic acid is = $\left(\frac{z-y}{x-y}\right)$ cal.

IV. TO FIND HEAT OF FORMATION OF MgO CALORIMETRICALLY

Chemicals and Apparatus

Mg powder, 0.5 (M) HCl, calorimeter set up.

Theory

The thermochemical reactions involved are:

(1) $Mg\ (s) + 2H^+\ (aq.) = Mg^{2+}\ (aq.) + H_2\ (g)$ ΔH_1

(2) $MgO\ (s) + 2H^+\ (aq.) = Mg^{2+}\ (aq.) + H_2O\ (l)$ ΔH_2

(3) $\frac{1}{2}\ O_2\ (g) + H_2\ (g) = H_2O\ (l)$ ΔH_3

(4) $Mg\ (s) + \frac{1}{2}\ O_2\ (g) = MgO\ (s)$ ΔH_4

Thus, $\Delta H_4 = \Delta H_1 - \Delta H_2 + \Delta H_3$ **(2.2)**

The standard value of ΔH_3 at 298 K is -68.32 kcal/mole. By measuring the values of ΔH_1 and ΔH_2 we can calculate the heat of formation of MgO (ΔH_4).

Procedure

1) Determine the water equivalent of the calorimeter as described in Expt. I.

2) Weigh 0.4 g of Mg metal powder.

3) Prepare a standard solution of HCl of strength 0.5 (M) as described in Experiment III (Table **2.5, 2.6** and **2.7**).

4) Take 200 ml of 0.5 (M) HCl in the calorimeter and record the temperature with time.

5) Add Mg metal powder to this acid and stir gently.

6) Record variation in temperature with time even after Mg has dissolved. Now calculate the heat of reaction ΔH_1 (Table **2.8**).

7) Weigh 0.705 g of MgO powder and dissolve it in 200 ml 0.5 (M) HCl and calculate the heat of reaction ΔH_2 in a similar manner as described in the last step.

Observation

Table 2.5. Preparation of (M/2) 250 ml oxalic acid solution.

The molecular weight of oxalic acid ($C_2O_4H_2.2H_2O$=126.03)

Initial Mass of Weighing Bottle (g)	Final Mass of Weighing Bottle (g)	Mass of Oxalic Acid Transferred (g)	Mass of Oxalic Acid to be Taken (g)	Strength of Oxalic Acid Solution (M)
-	-	-	15.75	-

Table 2.6. Standardization of NaOH against oxalic acid.

S. No.	Vol. of Oxalic Acid Taken (ml)	Volume of NaOH Consumed (ml)	Most Precise Reading (ml)
1	10	-	-

2		-	-
3		-	-

Table 2.7. Standardization of HCl against NaOH.

S. No.	Vol. of HCl Taken (ml)	Volume of NaOH Consumed (ml)	Most Precise Reading (ml)
1	10	-	-
2		-	-
3		-	-

Table 2.8. Calculation of heat of formation.

S. No.	ΔH_1 ($Mg + 2H^+$) cal	ΔH_2 ($MgO + 2H^+$)	ΔH_3 ($\frac{1}{2} O_2 + H_2$) cal	$\Delta H_3 = \Delta H_1 - \Delta H_2 + \Delta H_3$ cal
1	-	-	-	-
2	-	-	-	-
3	-	-	-	-

Result: The heat of formation of MgO is = …. Cal

CRYOSCOPY AND EBULLIOSCOPY

V. TO DETERMINE THE MOLAR MASS OF THE GIVEN SOLUTE IN WATER BY DEPRESSION IN FREEZING POINT METHOD

Chemicals and Apparatus

Solute (say sugar), NaCl (used in a freezing mixture), ice, Beckmann freezing point apparatus.

Theory

The molar mass of a solute is determined from the formula:

$$m_1 = \frac{1000K_f w_1}{w_2 \Delta T} \quad \textbf{(2.3)}$$

Where K_f is the freezing point constant. w_1 and w_2 are the weight of solute and solvent respectively. ΔT is depression at the freezing point.

Description of Beckmann Apparatus

The Beckmann apparatus consists of an inner tube A possessing a side tube that introduces the solute. The inner tube is fitted inside another tube A_1 which acts as an air jacket and ensures uniform and slower cooling of the liquid. The whole apparatus is capped in a glass jar B containing a freezing mixture. Beckmann thermometer stands a small stirrer (S) is dipped through two holes in a cork of inner tube. A larger stirrer S_1 is introduced in the outer vessel as shown in Fig. (**2.3**).

The Setting of Beckmann Thermometer

Beckmann thermometer is constructed for measuring out a small difference of temperatures at any point of the ordinary thermometer scale, but not the actual freezing points (or boiling point in specific cases). It consists of an open scale of only 5-6° graduated in 0.01°. It possesses a large bulb connected with an undulated glass tubing at the top which is closed at the upper end, with the help of a fine capillary glass tube having a uniform bore that runs over the porcelain scale. If we assume the zero of the thermometer is set at 20° C and that it is desired to set a new setting in such a way that 0° C falls on the upper part of the scale.

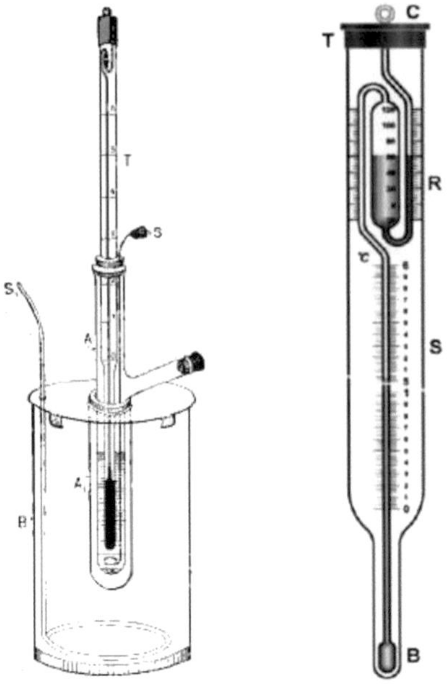

Fig. (2.3). Beckmann Apparatus and Beckmann thermometer.

In the beginning, the thermometer does not contain enough amount of mercury and if the bulb were immersed in ice water, the mercury column would disappear into the bulb. Tilt the thermometer to transfer the mercury from the reservoir to the space above the fine capillary in the seal. If the thermometer is kept in the upright position, the mercury can be frequently caused to run down from the space and join the main column by a quick jerk. If this does not take place readily, the bulb of the thermometer should be held in warm water, till mercury in the main column rises to join the mercury at the top. On cooling, the mercury will go down.

Now the bulb of the thermometer is placed in a beaker containing ice. The excess mercury in the top space can return to the reservoir by inverting the thermometer and giving it a sharp tap or jerk. This process should be carried out quickly before the mercury in the bulb become appreciable warmer. On again keeping the thermometer to ice bath the number of mercury drop increases in the thermometer. The excess amount of mercury should be removed cautiously by heating the bulb with warm water or even with the palm and forcing the excess mercury dropwise out of the capillary in the seal.

If we desire to set a thermometer to a temperature above that of the initial setting, the procedure will be similar.

For setting the Beckmann thermometer its bulb is placed in a beaker containing ice and then it is observed whether the mercury level is stationary on the thermometer scale or not. Stationarity indicates the proper setting of the thermometer. If a stationary condition is not achieved and the mercury level is much below then, it is required to add mercury from the upper reservoir. For this purpose, mercury thread is broken near the top by giving a sharp tap when the temperature of the bath is slightly higher than the freezing point of the pure solvent.

If mercury level is above the scale then it is required to transfer to the upper reservoir. For this purpose, the thermometer is placed in hot water and mercury is expelled until its amount is so adjusted that the mercury level stays on the scale.

Procedure

1) Beckmann thermometer is set following the procedure described above.

2) The whole apparatus is fitted as shown in Fig. **(2.3)**.

3) 20 ml of water is taken in an inner tube (A) and the Beckmann thermometer is immersed in such a manner that its bulb dips in the liquid. The liquid is stirred gently with the stirrer S_1 and allowed to super cool a bit below its freezing point.

4) Stir it vigorously when crystallization of ice starts. As the freezing initiates the thread of mercury begins to rise till it becomes stationary at a particular level. This reading gives a freezing point of the pure solvent.

5) The inner tube is removed and warmed to melt the solid solvent.

6) A definite weight of solute is added through the side tube. It is mixed thoroughly to obtain a homogeneous solution.

7) The freezing point of the solution is recorded (Table **2.9**) following steps 1- 4.

Observation

Weight of solvent (w_2) = ... g

The freezing point of pure solvent (T° C) = ...

Table 2.9. Depression in freezing point of the solvent.

S. No	Amount of Solute (w_1) g	Freezing Point of Solution (T_1)° C	Depression in Freezing Point ($\Delta T = T - T_1$)° C	Molecular Weight
	-	-	-	-
	-	-	-	-
	-	-	-	-

Calculations

The molar mass (m_1) of the solute = $\frac{1000 K_f w_1}{w_2 \Delta T}$

Result

Molar mass of the given solute = ……

Precautions

1) The temperature of the freezing mixture should be only 5° C below the freezing point of the pure solvent.

2) The reading of the freezing point must be noted as soon as crystallization starts, *i.e.* when the level of mercury becomes fixed.

3) Since, the ionic salts, like NaCl are formed by the electrostatic interactions, hence their formula weight is not very accurate. Hence finding molar mass by this experiment may give an erroneous result. Hence it is better to avoid finding the molar mass of ionic salts by this method.

VI. TO DETERMINE MOLAR MASS OF GIVEN SOLUTE IN WATER BY ELEVATION OF BOILING POINT METHOD

Chemicals and Apparatus

Given solute (say sugar), water, Landsberger boiling point apparatus and steam generating flask, Beckmann thermometer, or any other accurate thermometer.

Theory

The molar mass of the solute is given by the formula

$$m_1 = \frac{1000K_b w_1}{w_2 \Delta T} \quad \textbf{(2.4)}$$

K_b is molal elevation constant. Other symbols have already been mentioned in the earlier experiment.

Procedure

1) Set the Beckmann thermometer as explained in the previous experiment. However, for practical purposes, we can use any other accurate thermometer.
2) Landsberger type boiling point apparatus is used to perform the experiment the rose head ensures uniform distribution of steam through the liquid.
3) A known quantity (say 20 ml) of water is taken in the inner graduated tube. The thermometer is adjusted in such a manner that its bulb lies about 1 cm above the level of water.
4) Water is boiled in the vessel and steam is passed through a delivery tube.
5) The temperature (boiling point of pure water) is recorded when it becomes constant.
6) Take out the tube and dissolve a weighed quantity of the given solute in water.
7) Boiling temperature of this solution is recorded as described in step 5.
8) Record another set of readings by dissolving 0.5 g more solute and record the boiling point of the resulting solution (Table **2.10**).

Observations

Weight of solvent (w_2) = …. g

The boiling point of pure solvent (T° C) = ….

Table 2.10. Elevation in the boiling point of the solvent.

S. No	Amount of Solute (w_1) g	Boiling Point of Solution (T_1)° C	Elevation in Boiling Point ($\Delta T = T_1 - T$)° C	Molecular Weight
-	-	-	-	-
-	-	-	-	-
-	-	-	-	-

Calculations

The molecular weight (m_1) of the solute $= \frac{1000 K_f w_1}{w_2 \Delta T}$

Result

Molar mass of the given solute = ……

Precautions

1) The relation holds good for dilute solutions so take a small quantity of solute.

2) The rose head should be used for uniform heating of the liquid.

3) Since, the ionic salts, like NaCl are formed by the electrostatic interactions, hence their formula weight is not very accurate. Hence finding molar mass by this experiment may give an erroneous result. Hence it is better to avoid finding the molar mass of ionic salts by this method.

FURTHER READING

[1] Schulte M.; Thermochemistry. In: Geochemistry. Encyclopedia of Earth Science. Springer, Dordrecht. 1998. [http://dx.doi.org/10.1007/1-4020-4496-8_320]

[2] West, C. J.; New books: the thermochemistry of the chemical substances. *Journal of Physical Chemistry* 1937 41 (2), 333-333. [http://dx.doi.org/10.1021/j150380a014]

CHAPTER 3

Chemical Kinetics Experiments

INTRODUCTORY REMARKS

As the chemical reaction proceeds the concentration (amount) of reactant decreases while the product is formed. For a particular reaction, the change of concentration of reactant or product with respect to time is known as the rate of the reaction. Thus, in terms of mathematical representation, we can represent the rate of a chemical reaction as: $-\frac{dc}{dt}$. Chemical kinetics is a branch of physical chemistry that deals with the rate of a chemical reaction and its mechanism. The general form of rate of a chemical reaction can be represented in terms of concentration of reactant:

$$-\frac{dc}{dt} \propto (C_{Reactant_1})^{x^1}(C_{Reactant_2})^{x^2}....(C_{Reactant_n})^{x^y}$$

$$\text{where, } x^1+x^2+...+x^y=n \qquad \textbf{(3.1)}$$

Here k and n are the rate constant and order for the reaction. Order of reaction can have non-negative integer values as well as fractional values. Every reaction has unique order. Based on the order of reaction we can classify it as zeroth order, first order, second order, *etc*.

The general expression of the unit of the rate constant of a reaction can be expressed as $(\text{concentration})^{1-n}\times\text{time}^{-1}$. Thus, if the concentration is expressed in terms of mole/lit, then the rate constant can be expressed as $(\text{mole})^{n-1}\times(\text{lit})^{1-n}\times(\text{sec})^{-1}$. The rate of reaction is expressed as concentration/time.

> Always keep in mind that, the reaction rate is very much dependent on time, concentration and temperature. So, measure accurately time interval, concentration and never forget to read and note temperature.

Zero Order Reaction

In the case of a zero-order reaction, the rate is entirely independent of the concentration of reactants. For example, for any reaction A→B If the rate is independent of the concentration of A, then we can say that the reaction is obeying zero order in terms of reactant.

Charu Arora and Sumantra Bhattacharya

Zero-order reactions are often referred to as pseudo-zero-order reactions because the zero-order kinetics cannot continue when the reactant gets completely consumed.

Mathematically we can express the rate of a reaction as:

$$-\frac{dc}{dt}=k \qquad \textbf{(3.2)}$$

Example:

Decomposition of nitrous oxide at ~575° C. Hot Platinum wire plays a catalytic role. If a platinum wire is not applied, it follows second-order kinetics.

$$2N_2O \rightarrow 2N_2+O_2$$

First Order Reaction

In the case of a first-order reaction, the rate is dependent on the concentration of reactants. For example, for any reaction A→B if the rate is dependent on the concentration of A, then we can say that the reaction is obeying the first order in terms of reactant.

Example:

There are huge numbers of reactions, which follow first-order kinetics. Among them, dissociation of ethyl bromide:

$$CH_3CH_2Br \rightarrow C_2H_4 + HBr$$

Mathematically, we can express the rate of a first-order reaction (A→ Product), following first-order kinetics as:

$$-\frac{dc}{dt}=kc_A$$

$$-\frac{dc}{c_A} = kdt \qquad \textbf{(3.3)}$$

Upon integration, Equation (3.3) becomes:

$$-\int \frac{dc_A}{c_A} = k\int dt$$

On integration,

$$\ln c_{A_0} - \ln c_{A_t} = kt$$

$$\ln\left(\frac{c_{A_0}}{c_{A_t}}\right)=kt \quad (3.4)$$

Where c_0, c_t is the concentrations of the reactant at time 0 and time t, respectively. Thus, the unit of rate constant for a first-order reaction is time^{-1}. Graphically, the rate constant of the first-order reaction can be measured by plotting time *versus* the logarithm of concentration. A straight line is obtained for the first-order reaction, where c_0 is the intercept of the concentration axis and the slope is equal to the rate constant.

Second-Order Reaction

In the case of a second-order reaction, the rate is dependent on the product of two concentrations of reactants. For example, for any reaction A + B → product.

Example: Example of second-order reactions is base-catalysed hydrolysis of ester, esterification of acids, *e.g.*

$$CH_3COOC_2H_5+^{-}OH \rightarrow CH_3COO^{-}+C_2H_5OH$$
$$CH_3COOH+CH_3OH \rightarrow CH_3COOCH_3+H_2O$$

The formation of hydrogen iodide also follows second order kinetics:

$$H_2 + I_2 \rightarrow 2HI$$

General mathematical expression for the rate of second-order (A + B → product) reaction is: $-\frac{dc}{dt}=kc_Ac_B$, where c_A and c_B are the concentration of A and B, respectively. Two cases can arise here:

1) When the concentrations of A and B are different.

2) When the concentration of A and B are the same.

For the first case the rate constant can be expressed as:

$$k=\frac{1}{t(a_0-b_0)}\ln\left(\frac{b_0a_t}{a_0b_t}\right)$$

$$\text{or, } kt=\frac{1}{(a_0-b_0)}\ln\left(\frac{b_0a_t}{a_0b_t}\right) \quad (3.5)$$

Where, a_0 and b_0 are the initial concentrations of reactant A and B, respectively, and a_t, b_t the remaining concentrations of reactants after time t. The same expression for case 2 is:

$$k=\frac{a_0-a_t}{a_0a_tt}$$

$$\text{or, } kt=\frac{1}{a_t}-\frac{1}{a_0} \quad \textbf{(3.6)}$$

Any physical chemistry book can be consulted for derivation. Like the determination of the rate constant of first-order reactions, that of second-order reactions can also be determined graphically. For the different reactants, the plot of t *vs.* $\ln\,(b_0a_t/a_0b_t)\ln\left(b_0a_t/a_0b_t\right)$ is a straight line, passing through the origin and the rate constant is $k=\frac{\text{slope}}{(a_0-b_0)}$. On the other hand, for the second case, the plot of t *versus* the inverse of concentration at time t is a straight line with an intercept at Y-axis being inverse of initial concentration and the rate constant can directly be measured from the slope.

I. TO STUDY THE REACTION BETWEEN ACETONE AND IODINE IN PRESENCE OF ACID

Theory

In this laboratory work, the rate law describing the reaction between iodine and acetone in acidic solution will be determined experimentally using the initial rate method. To completely determine the rate law, the order of the reaction with respect to each reagent and the rate constant at the reaction temperature must be determined.

In the study of reaction kinetics, the rate at which a chemical reaction occurs depends on several factors, which are as follows:

1) Chemical properties of the reactants.
2) Concentrations of the reactants.
3) Reaction temperature.
4) Presence of catalysts (if any).

The rate of a reaction at a particular temperature is generally expressed in a mathematical equation of the form:

$$rate=kc_A^m c_B^n \tag{3.7}$$

In the rate law above, k is the rate constant; [A] and [B] are the molar concentrations of reagents A and B; and m and n are the orders of the reaction with respect to reagents A and B. The effects of temperature and catalysts are conveyed in the numerical value of the rate constant, k. The reaction being studied is that between iodine and acetone (as shown below), which occurs in an aqueous solution in the presence of HCl:

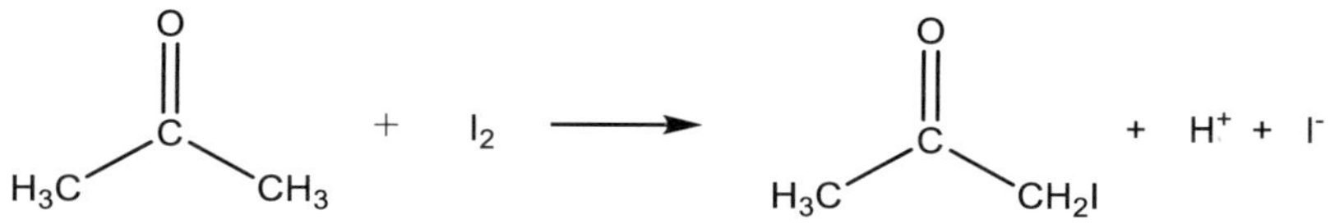

Although H^+ does not appear as a reagent in this reaction, experimental observation shows that the reaction does not proceed unless H^+ is present in the initial mixture. Therefore, the order of the reaction will be determined with respect to iodine, acetone, and H^+ concentrations.

Iodine has a characteristic brown colour in an aqueous solution. The rate of the reaction can therefore be followed by measuring the time it takes for all the iodine initially present to react- the time it takes for the solution to convert from coloured to clear and colourless. The experimental rate for each run is calculated as:

$$rate=\frac{\Delta I_2}{\Delta t} \tag{3.8}$$

The procedure here allows student groups to choose the specific concentrations of reagents to be used. To get reliable results, it is imperative that the scientific method be systematically employed. The scientific method assumes that variables are controlled and that only one variable at a time is changed. Therefore, as the initial concentration of components is varied to determine the rate law, only one of the reagent concentrations should be changed between any pair of runs.

The use of a systematic approach will allow easy determination of the effect of a specific change in concentration of a particular reactant on the rate. To simplify the calculations, consider changing the concentration of a reactant from one trial to the next by an integer factor (doubling, tripling, halving, *etc.*). This will make the subsequent mathematical analysis a lot easier than changing a concentration by a factor of 2.46, for example.

The Objectives for this experiment are to:
1) Design an initial rates procedure applying the scientific method.
2) Determine the order of the reaction with respect to iodine, acetone, and HCl.
3) Calculate rate constant (k), the rate constant, at lab temperature and graphically.
4) Examine the effect of temperature on rate constant (k).
The rate of the reaction can also be expressed as the change in the concentration of a reactant divided by the time interval:rate=$\frac{\Delta I_2}{\Delta t}$.

The iodination of acetone is easily investigated because iodine (I_2) has a deep yellow/brown color. As the acetone is iodinated and the iodine converted to the iodide anion, this color will disappear, allowing the rate of the reaction to be easily monitored.

We can study the rate of this reaction by simply making I_2 the limiting reaction a large excess of acetone and H^+ ion. By measuring the time required for the initial concentration of iodine (I_2) to be used up completely, the rate of the reaction can be determined by the equation.

$$\text{rate}=\frac{\Delta I_2}{\Delta t}=\frac{(c_{I_2})_{\text{final}}-(c_{I_2})_{\text{initial}}}{t_{\text{final}}\text{-}t_{\text{initial}}}=\frac{(c_{I_2})_{\text{initial}}}{t_{\text{final}}} \quad \textbf{(3.9)}$$

Thus, the rate is a change in iodine concentration per unit time. From the rate information, we can determine the orders with respect to acetone (m), acid (n), and iodine (p) by varying the amounts of reactants and measuring the effect on the rate. Once the orders of reaction are known, we will be able to calculate the rate constant, k. In Part One of this experiment, you will determine the rates of reactions, the orders of the reactants, and finally the rate constant at room temperature.

Procedure

1) Take a couple of Conical flasks, one containing 10 ml acetone, 20 ml 0.5 (M) sulphuric acid, and 60 ml of distilled water and another containing 50 ml of iodine solution only are kept in a thermostat at 25°C. They are allowed to attain equilibrium temperature.

2) 10 ml of 0.1 (M) iodine solution is added to the flask containing acetone, sulphuric acid, and distilled water (Table **3.1**). Stopwatch is started when the pipette is half discharged. The solution is thoroughly mixed and 10 ml of solution is taken in a conical flask. 10 ml of 1 (M) sodium acetate solution is added to it to check the reaction. The reaction can also be rechecked with the help of sodium bicarbonate.

3) The reaction mixture is titrated with 0.1 (M) sodium thiosulphate (known as a hypo) using starch as an indicator. The residual iodine can be determined with the help of titre value. in 5 min, 10 ml of the reaction mixture is withdrawn and followed by proceeding similarly. This process is repeated (Table **3.2**) with varying amounts of acetone, iodine, and sulphuric acid to explore the effect of changes in their concentrations.

Observations

Experimental Temperature = ... ° C

Table 3.1. Preparation of Solutions for the Experiment.

Bottle no.	Acetone (ml)	(M/2) H_2SO_4 (ml)	(N/10) I_2 solution (ml)	Water (ml)
1	10	20	10	60
2	10	20	5	65
3	5	20	10	65
4	10	10	10	70

Table 3.2. Titration of Solutions Prepared Against Hypo Solution.

Bottle no. 1		Bottle no. 2		Bottle no. 3		Bottle no. 4	
Time (min)	Vol of hypo (ml)	Time (min)	Vol of hypo (ml)	Time (min)	Vol of hypo (ml)	Time (min)	Vol of hypo (ml)

Calculations

A plot of time *vs* titre value for all the bottles and find the slope for each curve. The slope gives the value of the rate constant, k. In this case, a straight line (parallel to the time axis, Fig. (**3.1**) is observed, indicating the reaction to be zero-order with respect to iodine.

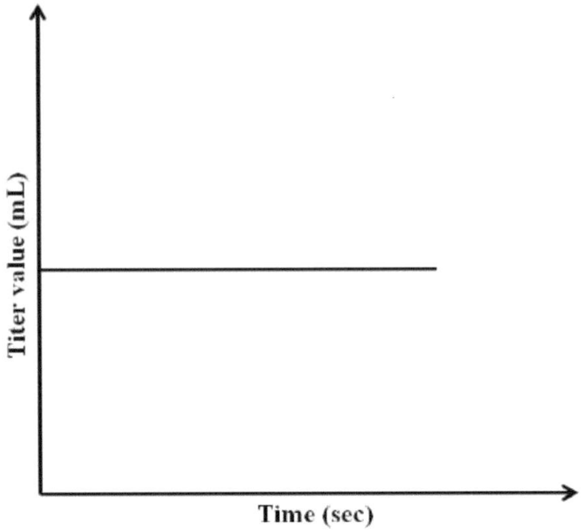

Fig. (3.1). A Plot of Time *vs.* titer value.

For bottle no. 1 and 2 from the values of rate constant, the order of reaction with respect to acetone is calculated as follows:

When acetone and acid are taken in excess: $-\frac{dc_{I_2}}{dt} = k_1 c_{I_2}^x$ **(3.10)**

$$k_1 = kc_{acetone_1}^{y} c_{acid}^{z} \quad \textbf{(3.11)}$$

Bottle 1 and 3 (when acetone concentration is halved) the new value of k_1 is given by:

$$k_{1'} = kc_{acetone_2}^{y} c_{acid}^{z} \quad \textbf{(3.12)}$$

From Equation (3.11) and Equation (3.12), we have:

$$\frac{k_1}{k_{1'}} = \left(\frac{c_{acetone_1}}{c_{acetone_2}}\right)^y = 2^y \quad \textbf{(3.13)}$$

Similarly, for the set of bottles 1 and 4, we can calculate the value of z using the values of k in these cases to determine the order with respect to acid.

Result

The order of reaction with respect to iodine, acetone, and acid is.... .... at ...° C.

Precautions

1) Be careful in using acetone. It is hazardous to health. So do not keep the acetone bottle opened, as this prevents acetone evaporation.
2) Iodine solution should be prepared in a minimum amount of potassium iodide.

II. A. TO FIND RATE CONSTANT OF HYDROLYSIS OF METHYL ACETATE CATALYSED BY AN ACID AND DETERMINATION OF HALF-LIFE OF THE REACTION

Chemicals and Equipment Required

Methyl acetate (MA) (freshly distilled), (M/2) HCl, 0.05 (M) NaOH, phenolphthalein indicator, water bath or thermostat, stopwatch, beakers, conical flask with stopper, burettes, pipettes (25 ml, 5 ml, and 2 ml)

Theory

Acid acts as a catalyst for hydrolysis of methyl acetate

$$MA + H_2O + H^+ \rightarrow AcOH + MeOH + H^+$$

Where, Ac and Me are acetyl and methyl groups, respectively. The rate of reaction is given by:

$$\frac{dx}{dt} = kc_{MA}C_{H_2O}$$

As water is present in excess, its concentration practically remains physically constant throughout the reaction. So, rate expression becomes

$$\frac{dx}{dt} = kc_{MA} \qquad \textbf{(3.14)}$$

It is a pseudo-first-order reaction as the rate of a reaction is determined by the first power of the concentration of the ester. The acetic acid formed during the process can be estimated by titrating the reaction mixture with the standard solution of an alkali. As the reaction follows first-order kinetics, the half-life period of the reaction is given by:

$$t_{1/2} = 0.693/k \quad \textbf{(3.15)}$$

Procedure

1) Methyl acetate (freshly distilled) and (M/2) HCl are kept in a water bath for about half an hour in separate bottles. After acquiring the temperature of the bath, 2 ml of methyl acetate is added into a conical flask, containing 50 ml of (M/2) HCl.

2) Stopwatch is started when half the pipette has been discharged. Shake the mixture and withdraw a 2 ml reaction mixture using the 2 ml pipette and transfer it to a flask containing 25 ml ice-cold water to arrest the reaction.

3) The solution is titrated against (M/20) NaOH, using phenolphthalein indicator. The solution is stirred during titration and the endpoint is recorded at the appearance of light pink colour. A similar experiment is performed for 45 minutes with an interval of 5 min (Table **3.3**).

4) For determination of V_{inf}, *i.e.*, when hydrolysis of the ester is complete, 25 ml of the reaction mixture is transferred in a separate conical flask and kept in a water bath at 60-70° C for half an hour. The flask is allowed to cool to room temperature.

5) 2 ml reaction mixture is titrated as described above to get the value of V_{inf}. The amount of NaOH used is equivalent to the total amount of HCl initially present and the amount of acetic acid formed during the reaction. An initial amount of HCl present can be determined by titrating against the same alkali before starting the reaction. Acetic acid produced after different intervals of time can be estimated by titration.

Observation

Temperature = ……° C

Table 3.3. Titration of reaction mixture against NaOH at different time.

Time (min)	Vol. of Reaction Mix (ml)	Vol. of NaOH Required (V_t) (ml)	(V_{inf}-V_t) ml
0	2	V_0	(V_{inf}-V_0)
5	2	V_5	(V_{inf}-V_5)
10	2	V_{10}	(V_{inf}-V_{10})
15	2	V_{15}	(V_{inf}-V_{15})
.....	2	....	.....
continue till 45 min	2	....	.....
Inf	2	....	.....

The amount of acetic acid generated at the end of the reaction is the same as the initial amount (a) of methyl acetate.

The amount of acetic acid produced after time t, *i.e.*

$$(a - x) \propto (V_{inf} - V_0) - (V_t - V_0)$$

$$\propto (V_{inf} - V_t) \quad \textbf{(3.16)}$$

Calculation

The value of k is now calculated following first-order rate expression

$$k = \frac{2.303}{t} \log\left(\frac{a}{a-x}\right) = \frac{2.303}{t} \log\left(\frac{(V_t\text{-}V_0)}{(V_{inf}\text{-}V_t)}\right) \quad \textbf{(3.17)}$$

The value of the rate constant can be calculated at different time intervals. The values of k at different times are found to be nearly constant.

The half-life of the reaction is determined by: $t_{1/2}$=0.693/k. $t_{1/2}$ can be calculated by substituting the value of k. t *vs* $\log\left(\frac{(V_t\text{-}V_0)}{(V_{inf}\text{-}V_t)}\right)$ is plotted (Fig. **3.2**). The slope of the straight line obtained gives the value of k **(Guggenheim method).**

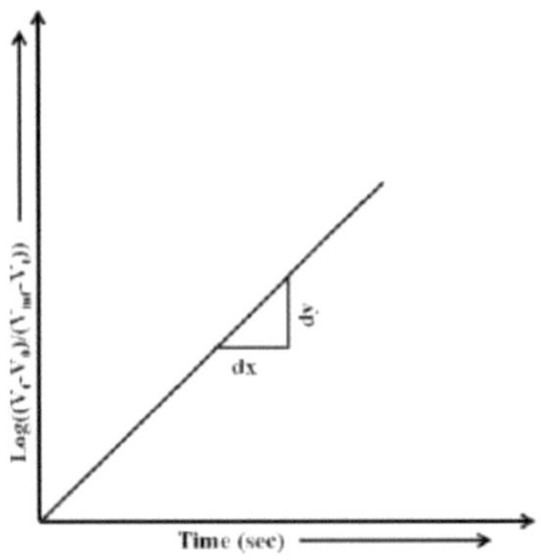

Fig. (3.2). Plot of $\log\left(\frac{(V_t\text{-}V_0)}{(V_{inf}\text{-}V_t)}\right)$ *vs.* time.

Results

The rate constant for acid-catalysed hydrolysis of methyl acetate = …... min^{-1}

The half-life of the reaction = …... min

Precautions

1) The temperature of the reaction mixture should be kept constant throughout the experiment.

2) Distilled water should be free from CO_2 to be used for preparing all the solutions. Ice should also be made of distilled water.

3) The zero time should be recorded when ice is added to the first 2 ml of the reaction mixture.

4) Be punctual and accurate about recording time and time intervals. Please note that precaution is common for every kinetics experiment and punctuality is an integral part of life.

II. B. 1st ORDER REACTION: TO DETERMINE TEMPERATURE COEFFICIENT OF ACID CATALYSED HYDROLYSIS OF METHYL ACETATE AND ITS ENERGY OF ACTIVATION

Experiment

To study the hydrolysis of methyl acetate in presence of hydrochloric acid and find its activation energy.

Chemicals and Equipment Required

0.1 (M) NaOH, methyl acetate, 0.5 (M) HCl, stopwatch, water bath, phenolphthalein, conical flask, beakers.

Theory

The reaction is catalysed by H^+ ions of an acid (HCl). This reaction is an example of pseudo-unimolecular reactions. Since water is present in large excess, its concentration is practically constant throughout the reaction. The concentration of HCl (catalyst) also remains constant. Therefore, the rate of reaction depends only upon the concentration of ester.

$$\text{rate}=\frac{dx}{dt}=kc_{AcOMe} \quad \textbf{(3.18)}$$

Hence, the reaction is the first order. During the hydrolysis of ester, acetic acid is produced. Therefore, the progress of the reaction is followed by determining the amount of acetic acid formed at different time intervals. A definite quantity of the reaction mixture is withdrawn after different time intervals and is titrated against a standard solution of alkali. The amount of alkali used is equivalent to the total amount of HCl present initially and the amount of acetic acid formed. The volume of alkali used at the start of the reaction is equivalent to the amount of HCl alone. Hence, the amount of acetic acid formed (x) after different intervals of time can be calculated. The amount of acetic acid formed at the end of the reaction is equivalent to the initial concentration of ester (a). Suppose the volumes of alkali used or required for the reaction at the start, after time t and the end of the reaction, are V_0 and V_t, respectively. The amount of acetic acid (x) produced after time t is directly proportional to $V_t - V_0$. The initial concentration of methyl acetate (a) is directly proportional to $V_{inf} - V_0$. Hence, the amount of ester present is (a-x) at time t.

$$(a\text{-}x) \propto \{(V_{inf}\text{-}V_0) - (V_t\text{-}V_0)\}$$
$$\text{or, } (a\text{-}x) \propto (V_{inf} - V_t) \quad \textbf{(3.19)}$$

The value of the rate constant (k) for first-order kinetics is:

$$k=\frac{2.303}{t}\log\left(\frac{a}{a\text{-}x}\right)$$

$$\text{or, } k=\frac{2.303}{t}\log\left(\frac{(V_{inf}\text{-}V_0)}{(V_{inf}-V_t)}\right) \quad \textbf{(3.20)}$$

The value of k comes out constant during different intervals of time, indicating the reaction to be first order.

Procedure

1) Take 50 ml of 0.5 (M) HCl in a clean dry, 250 ml conical flask and about 10 ml of pure methyl acetate in a test tube. Cork both of them and place them in a thermostat or water bath at or near room temperature.

2) Keep the 0.5 (M) HCl and methyl acetate in the thermostat or water bath for about 10 min to allow them to acquire the temperature of the bath.

3) In the meantime, fit the burette properly and fill it with 0.1N NaOH solution. Also, add 25 ml of ice-cold water in six conical flasks.

4) Pipette out 5 ml of the ester from the conical flask and add it to the flask containing 50 ml of 0.5 (M) HCl.

5) Shake the contents, pipette out 10 ml of the reaction mixture, and transfer it at once to a conical flask containing ice-cold water. Titrate it against 0.1 (N) NaOH taken in the burette by using phenolphthalein as an indicator. The pink color appears at the endpoint. The volume of 0.1 (N) NaOH used against the withdrawn sample of the ester and dilute HCl mixture is taken as V_0.

6) Pipette out 10 ml of the mixture and add it to the conical flask containing ice-cold water after 10 min. Titrate it against 0.1 (N) NaOH. This gives V_t after 10 min (Table **3.4**).

7) Repeat the above procedure for every 10 min interval. Place the remaining reaction mixture in a separate water bath at 60-70°C for about one hour. Pipette out

10 ml of the mixture and titrate it against alkali. Repeat the same experiment at 40° C and calculate E_a.

Results

Experimental temperature = ...° C.

Table 3.4. Progress of reaction with time.

Sl. No	Time (min)	Volume of NaOH (ml)
1	0	
2	10	
3	20	
4	30	
5	40	

8) Pipette out 5 ml of the ester from the conical flask and add it to the flask containing 50 ml of 0.5 (N) HCl. Shake the contents, pipette out 10 ml of the reaction mixture and transfer it at once to a conical flask containing ice-cold water. Titrate it against 0.1 N NaOH taken in burette by using phenolphthalein as an indicator. The appearance of pink color takes place at the endpoint. The volume of 0.1 N NaOH was used against the withdrawn sample of the ester and the dilute HCl mixture is taken as V_0. Pipette out 10 ml of the mixture and add it to the conical flask containing ice-cold water after 10 min. Titrate it against 0.1 (N) NaOH. This gives V_t after 10 min.

9) Repeat the above procedure every 10 min interval. Place the remaining reaction mixture in a separate water bath at 60-70° C for about one hour. Pipette out 10 ml of the mixture and titrate it against alkali. Repeat the same experiment at 40°C and calculate E_a.

The plot of time of the progress of reaction *versus* (*vs*) log (V_{inf}-V_t). From the slope, calculate k_1 and k_2. E_a can be calculated using the following relationship:

$$\ln\left(\frac{k_1}{k_2}\right) = \frac{E_a}{R}\left(\frac{1}{T_2} - \frac{1}{T_1}\right)$$

$$\log\left(\frac{k_1}{k_2}\right) = \frac{E_a}{2.303R}\left(\frac{1}{T_2} - \frac{1}{T_1}\right) \tag{3.21}$$

Precautions

1) Don't drop the beakers from the thermostat.

2) Use ice-cold water only.

3) Perform the titrations properly.

4) Always take alkali in the burette.

III. TO STUDY THE KINETICS OF SAPONIFICATION OF ETHYL ACETATE WITH SODIUM HYDROXIDE

Chemicals and Apparatus Required

(M/40) strength of ethyl acetate (EA), HCl and NaOH, burette, pipette, 12 conical flasks (8 pieces of 100 ml and the rest are of 200 ml).

Theory

Base catalysed hydrolysis of organic acid is known as saponification. Soap is a Na or K salt of long-chain fatty acids. Thus, ester forms by this reaction. The reaction between EA and alkali takes place as follows:

$$EA + NaOH \rightarrow AcONa + EtOH$$

Where AcOH is acetic acid and EtOH is ethanol. The rate of reaction is given by:

$$\frac{dx}{dt} = kc_{EA}c_{-OH} \tag{3.22}$$

During this reaction, the concentration of alkali also changes with that of ester. Hence the rate of reaction is dependent on the concentration of both the reactants and the reaction follows second order kinetics. In acid catalysed hydrolysis by EA, the concentration of H^+ ions remain unchanged, so the reaction follows a first-order reaction with respect to the reactant. The base-catalysed hydrolysis is quite fast, so reactants may be taken in a concentration of (M/40) or even less.

Procedure

1) 50 ml of (M/40) EA and 100 ml (M/40) NaOH are kept in a water bath in two separate flasks to attain the same temperature.

2) Meanwhile, eight conical flasks (100 ml each) containing 50 ml ice-cold (M/40) HCl are arranged.

3) When the flask containing ester and alkali has acquired the same temperature, the alkali is poured into the ester solution rapidly. Note the time, when half volume of NaOH is discharged into EA. This is considered as 0 time. Immediately transfer 10 ml of reaction mixture into a conical flask containing 25 ml ice-cold (M/40) HCl. The time required for half discharged of the reaction mixture is considered as the time of stopping the reaction.

4) The un-reacted acid is estimated by titrating back by using standard alkali solution. Within the same time interval (say, 5 min) repeat the experiment for up to 1 hour. Let the volume of alkali required for titration at any time, t, be V_t. This value will increase with a gradual decrease in the concentration of NaOH in the reaction mixture. The infinite reading may be recorded either after 24 hours or after heating the reaction mixture in a thermostat at 70° C for about 30 min. The titre value V_{inf} is was determined as described earlier (Table **3.5**).

5) Suppose the initial concentration of NaOH be **a** and that of ester be **b**. The value of **a** can be calculated in 10 ml of the reaction mixture by applying the normality equation for HCl and NaOH. The value of V_0 can be determined by adding 5 ml alkali to 10 ml HCl and titrating it against the same solution of alkali ((M/40) NaOH). Similarly, 25 ml (M/40) HCl is titrated against (M/40) NaOH. Let the titre value be V. Now a ≡ (V- V_0) of (M/40) NaOH.

Observations

The titre value when 25 ml HCl is titrated against (M/40) NaOH = ... (V) ml

Experimental temperature = ... ° C

Table 3.5 Titration Value at Different Time Interval.

Time (min)	Vol. of NaOH Required (V_t ml)	a-x=V- V_t (ml)	b-x= V_{inf} -V_t (ml)
0	V_0		
5	V_5		
10	V_{10}		
…..	….		
Completion of reaction	V_{inf}		

Calculation

For second-order kinetics value of the rate constant can be expressed as follows:

$$k=\frac{2.303}{t(a-b)}\log\left(\frac{b(a-x)}{a(b-x)}\right) \qquad \textbf{(3.23)}$$

Here, a and b represent the initial concentration of NaOH and EA in 10 ml of the reaction mixture. The values of a and b can be calculated in terms of volume of (M/40) alkali as follows:

a-x is the amount of NaOH present in 10 ml of the reaction mixture at time t. This can be estimated with the help of the amount of HCl used from 25 ml of acid. This is equivalent to the difference between the initial amount of acid present and the amount of acid left at time t.

$$a-x \equiv (V-V_t)\text{ ml. So, } x=a-(a-x) = (V_t-V_0)\text{ ml}$$

(a-b) = Excess of NaOH over ester ≡ Amount of NaOH unused after completion of reaction ≡ $(V-V_{inf})$

So, $$b = a-(a-b) \equiv (V_{inf}-V_o)\text{ ml}$$

Thus, $$b-x = (V_{inf}-V_o)-(Vt-V_o) \equiv (V_{inf}-V_t)$$

Thus,

$$k=\frac{2.303}{t(a\text{-}b)}\log\left((V_{inf}-V_0)(V-V_t)/(V-V_0)(V_{inf}-V_t)\right) \qquad \textbf{(3.24)}$$

The value of (a-b) is expressed in terms of moles.lit^{-1} of the reaction mixture. In the present experiment a = b = 0.025.

The value of the rate constant, k, comes out to be constant at different time intervals, indicating the reaction to be second order.

Plot of $\log\left(\frac{(V_{inf}-V_0)(V-V_t)}{(V-V_0)(V_{inf}-V_t)}\right)$ as ordinate *vs* time in abscissa should be a straight line passing through the origin. From the slope of the graph, the rate constant can be determined (Fig. **3.3**).

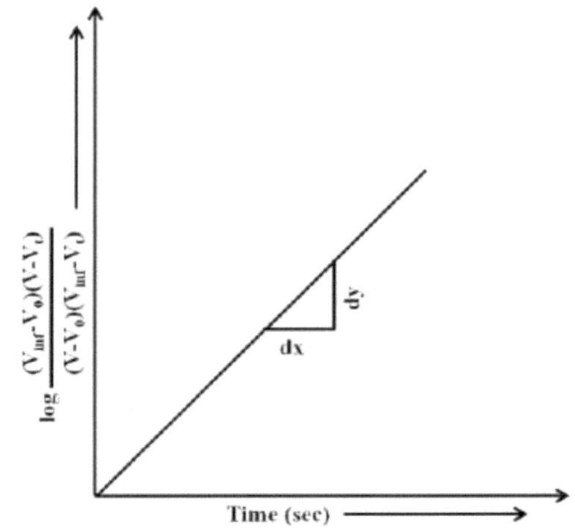

Fig. (3.3). Plot of $\log\left(\frac{(V_{inf}-V_0)(V-V_t)}{(V-V_0)(V_{inf}-V_t)}\right)$ *vs.* time.

Result

Value of rate constant determined from calculation = …… moles.lit^{-1}min^{-1}

Value of rate constant determined from plot = …… moles.lit^{-1}min^{-1}

Precaution

1) Handle carefully all the chemicals because all of them are injurious to health.

2) Temperature must remain constant throughout the experiment.

IV. TO FIND OUT VELOCITY CONSTANT AND ACTIVATION ENERGY OF REACTION BETWEEN POTASSIUM PERSULPHATE AND POTASSIUM IODIDE ALSO STUDY INFLUENCE OF IONIC STRENGTH AND RATE CONSTANT-AN EXAMPLE OF SECOND-ORDER REACTION

Chemicals and Apparatus Required

0.1 (M) potassium persulphate ($K_2S_2O_8$), 0.1 (M) potassium iodide (KI), thermostat, reagent bottles, ice, burette, pipette, conical flask.

Theory

The reaction between $K_2S_2O_8$ and KI takes place as follows:

$$K_2 S_2O_8 + 2KI = 2K_2SO_4 + I_2$$

The reaction takes place in two steps:

$$\text{(i) } S_2O_8^{2-} + I^- \rightarrow (S_2O_8I)^{3-} \text{ (slow)}$$

$$\text{(ii) } (S_2O_8I)^{3-} + I^- \rightarrow 2SO_4^{2-} + I^- \text{ (fast)}$$

If a and b represent the initial concentrations of $K_2S_2O_8$ and KI in the reaction mixture and x represent the amount of $K_2S_2O_8$ dissociated at equilibrium, then the rate of reaction can be expressed as:

$$\frac{dx}{dt} = kc_{S_2O_8^{2-}} c_{I^-} = k \times (b\text{-}2x) \times (a\text{-}x). \qquad \textbf{(3.25)}$$

Since the slowest step is the rate-determining step (RDS) therefore, the first step gives the expression for the rate equation.

The above equation indicates the reaction to be second order. On integrating the above equation, we get:

$$k = \frac{1}{2at}\frac{x}{a\text{-}x} \text{ (when b=2a)} \qquad \textbf{(3.26)}$$

Iodine is liberated during this reaction and the rate of reaction can be studied by titrating the liberated iodine in 5 ml (any certain amount) of reaction mixture against standard hypo solution at different time intervals. The iodine liberated can be estimated with the help of titre values, which is proportional to the amount of persulphate that disappeared during reaction from 5 ml of the reaction mixture. In this manner, we can find values of x at different time intervals. According to the Brønsted-Bjerrum equation for dilute solution:

$$\log k = \log k_0 + z_A z_B \sqrt{\mu} \quad \textbf{(3.27)}$$

Where, k is the rate constant at ionic strength, μ, k_0 is the rate constant at μ=0, the valences of ions for reactants are z_A and z_B.

Using Lewis and Randall expression μ can be determined as:

$$\mu = \frac{1}{2} \sum_i c_i z_i^2 \quad \textbf{(3.28)}$$

Here c_i is the concentration of ith reacting species.

The activation energy E can be estimated applying the Arrhenius equation:

$$k = Ae^{-\left(\frac{E}{RT}\right)} \quad \textbf{(3.29)}$$

Here A is the frequency factor.

On taking log, we get:

$$\ln k = \ln A - \frac{E}{RT} \quad \textbf{(3.30)}$$

On differentiating Equation (3.30) with respect to (wrt) temperature (T):

$$d(\ln k) = \frac{E}{RT^2} dT$$

$$\int_{k_1}^{k_2} d(\ln k) = \int_{T_1}^{T_2} \frac{E}{RT^2} dT$$

$$\text{or, } \ln\left(\frac{k_2}{k_1}\right) = \frac{E}{R}\left(\frac{1}{T_1} - \frac{1}{T_2}\right)$$

$$\text{or, } \ln\left(\frac{k_2}{k_1}\right) = \frac{E}{2.303R}\left(\frac{1}{T_1} - \frac{1}{T_2}\right) \qquad \textbf{(3.31)}$$

The value of activation energy E can be calculated after determining the values of k_1 and k_2 at two different temperatures T_1 and T_2.

Procedure

1) Prepare standard solutions of 0.1 (M) $K_2S_2O_8$ and 0.1 (M) KI.

2) Transfer 100 ml of 0.1 (M) $K_2S_2O_8$ and 100 ml 0.1 (M) KI into two reagent bottles separately and keep them in a water bath at a fixed temperature for half an hour and record the temperature of the water bath and add the KI solution to solution of $K_2S_2O_8$.

3) Start the stopwatch when half KI solution has been discharged to $K_2S_2O_8$ solution. Add some pieces of ice to a conical flask and pour 5 ml of reaction mixture into it. Titrate the reaction mixture with 0.1 (M) $Na_2S_2O_3$ solutions using a starch indicator.

4) In a similar manner, withdraw 5 ml of the reaction mixture after time intervals of 10, 20, 30, 40 min and titrate as described earlier (Table **3.6**). Increase the temperature of the water bath by 10° C and similarly repeat the procedure.

5) To study the effect of varying the ionic strength on the reaction, repeat the procedure by taking solutions of $K_2S_2O_8$ and KI of different strengths.

6) To estimate the titre value on completion of the reaction, pipette out 40 ml reaction mixture in a conical flask and add 4 g of KI and place the flask in a water bath at 60° C. The effect of increasing the temperature and concentration of KI will lead to quick completion of the reaction, liberating an equivalent amount of iodine. Allow the reaction mixture to cool down to room temperature and titrate 5 ml of the solution with 0.1 (M) $Na_2S_2O_3$ solution to obtain the titre value.

Observations

Temperature = ...° C

Initial concentration (a) = ... ml of 0.01 (M) hypo solution

Table 3.6. Titration value at different time interval.

Time (t) (min)	Titre Value (x) (ml)	(a-x)	1/(a-x)	$k=\frac{1}{t}\frac{x}{a(a-x)}$
5				
10				
15				
.....				
40				

Calculations

Initial concentrations of $K_2S_2O_8$ and KI in terms of 0.01 (M) hypo can be calculated by a simple experiment. Mix equal volumes of 0.1 (M) $K_2S_2O_8$ and 0.1 (M) KI solutions. Now the normality of $K_2S_2O_8$ is changed to 0.05 (M)in the mixture. This is equivalent to 25 ml of 0.01 (N) hypo solution. In the present experiment, the initial concentrations of reactants are the same. So, the rate constant can be expressed as $k=\frac{1}{t}\frac{x}{a(a-x)}$. Here, x is equivalent to titre value of 0.1 (N) hypo at time t (v_t). Therefore, $k=\frac{1}{t}\frac{v_t}{25(25-v_t)}$.

Plot the values of log k as ordinate (Y-axis) *vs* $\sqrt{\mu}$ along abscissa (X-axis), to study the influence of ionic strength on rate constant. A straight line with a slope equal to 2, *i.e*, $z_A z_B$ is obtained (Fig. **3.4**).

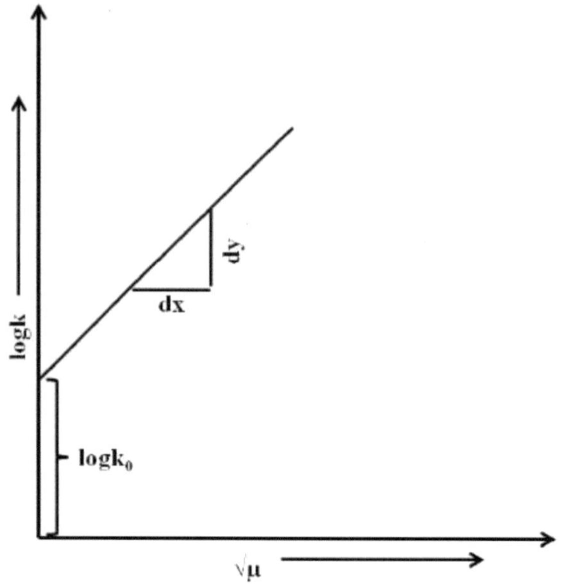

Fig. (3.4). Plot of log K *vs.* $\sqrt{\mu}$.

The energy of activation can be evaluated from the equation:

$$\log_{10}\left(\frac{k_2}{k_1}\right) = \frac{E}{2.303R}\left(\frac{1}{T_1} - \frac{1}{T_2}\right) \quad \textbf{(3.32)}$$

Result

The rate constant of the reaction is …… $mol^{-1}min^{-1}$

The value of activation energy is …... $cal.mol^{-1}$

Precautions

Iodine solution should be prepared in a minimum quantity of KI.

V. KINETICS OF IODINE CLOCK REACTION: (A) TO STUDY THE KINETICS OF IODINE CLOCK REACTION AND (B) TO DETERMINE THE RATE CONSTANT FOR OXIDATION OF IODIDE IONS BY H_2O_2, STUDYING THE KINETICS AS AN IODINE CLOCK REACTION

Chemicals and Equipment Required

250 ml of 0.01(M) potassium persulphate. Let us label it as A, 250 ml of 0.3(M)KI+0.0005 (M) hypo solution + starch solution, 10 ml in 250 ml of the mixed solution and mark as B with measuring cylinder, pipette, burette, and titration flask, neat and clean white paper, thermostat, thermometer, and stopwatch.

Theory

In the reaction of potassium persulphate with potassium iodide, containing starch and low molarity solution of sodium thiosulphate, the persulphate ions preferentially oxidize iodide ions than thiosulphate ions.

$$S_2O_8^{2-} + 2I^- \rightarrow 2SO_4^{2-} + I_2$$

The liberated iodine in this step does not change the starch blue as it reacts with thiosulphate ions immediately. This step is rate-determining.

$$I_2 + 2S_2O_3^{2-} \rightarrow S_4O_6^{2-} + 2I^-$$

The above reaction continues till the consumption of all the thiosulphate ions and then the solution turns blue, because of the reaction of iodine with starch. In this step, the thiosulphate acts as a monitor

$$I_2 + \text{starch} \rightarrow \text{blue colour}$$

The time required for the appearance of blue colour depends on two factors:

1) Initial rate of formation of iodine which depends on the concentration of iodide and persulphate ions.

2) Amount of thiosulphate present.

The time for the appearance of blue colour has a clock-like accuracy for particular conditions of temperature and concentration, that's why it is termed as clock reaction. The blue colour can be discharged by adding more thiosulphate, which reappears after some time. The process of reappearance of blue colour can be repeated a number of times by adding further quantities of thiosulphate, till the persulphate ions get exhausted. The subsequent time intervals will be longer, even with the equal addition of thiosulphate. As the concentration of persulphate ions continues to decrease, the reaction given in the first step turns slower and slower.

Procedure

1) Keep both solutions in a trough of water at constant temperature and swirl them from time to time until they acquired the same temperature as the water bath.

2) Check temperature of the water bath and pipette the 25 ml of the solution A in dry 100 ml of a beaker.

3) Pour 25 ml solution of B into A, start the stopwatch, mix the solution and place the beaker on white paper. Note the time when the blue colour first appears.

4) Repeat the mixing of the measured solution and note the time again when the blue colour reappears. The experiment is repeated by changing the concentration of potassium persulphate as shown in Table **3.7** below:

Table 3.7. Solution preparation.

Bottle no.	Vol. of Sol. A (ml)	Vol. of Sol. B (ml)	Vol. of Water (ml)
1	25	25	0
2	20	25	5
3	15	25	10
4	10	25	15

Temperature is maintained at 20°C and the initial rate method is used for analysing the experimental data.

Repeat the experiment with double concentration of the KI and hypo solution and interpret the result.

B. TO DETERMINE THE RATE CONSTANT FOR OXIDATION OF IODIDE IONS BY H_2O_2, STUDYING THE KINETICS AS AN IODINE CLOCK REACTION

Theory

Reactions may be classified according to the number of molecules taking part in the reaction, which determines their molecularity, or according to the number of molecules whose concentrations affect the rate of the reactions, which determines the order of the reaction. Hydrogen peroxide decomposes according to the following reaction:

$$H_2O_2 + 2H^+ + 2I^- \rightarrow 2H_2O + I_2 \text{ (slow)}$$

The iodine liberated in this step reacts instantaneously with thiosulphate ions, so it does not colour starch solution blue:

$$I_2 + 2S_2O_3^{2-} \rightarrow S_4O_6^{2-} + 2I^-$$

This reaction continues till the complete consumption of thiosulphate and then excess iodine turns starch solution blue. The time involved for the appearance of blue colour depends on the initial rate of formation of iodine which is dependent on the concentration of peroxide and iodide ions, amount of thiosulphate present in the reaction mixture.

The time required for the appearance of blue colour, for any specific concentration and temperature has clock-like accuracy. That's why the reaction is termed a clock reaction or iodine clock reaction. On adding the same quantity of thiosulphate, the blue colour gets discharged and reappears after some time. This process can be repeated many times by adding more quantity of thiosulphate, till the whole quantity of H_2O_2 is consumed. The subsequent time intervals will be longer even when the quantities of thiosulphate added are equal because the concentration of peroxide keeps on decreasing and the main reaction turns slower. The order of the reaction can be found out by taking large excess of iodide and hydrogen ions. As the rate equation is:

$$\text{rate} = kc_{H_2O_2}^{x} c_{I^-}^{y} c_{H^+}^{z} \quad \textbf{(3.33)}$$

Procedure

1) Take 150 ml of distilled water in a conical flask and add 20 ml 1 (M) KI solution, 10 ml of 2 (M) sulfuric acid, 1ml starch, and 5 ml 0.025 (M) hypo solution to it. This conical flask as well as another conical flask containing 0.1 (M) H_2O_2 is kept in a water bath. 5 ml of 0.1 (M) H_2O_2 solution is added with a pipette and record the time of addition. The reaction mixture is mixed thoroughly and kept in a water bath. Record the time involved for the appearance of blue colour, without stopping the stopwatch.

2) 5 ml of 0.025 (M) hypo solution is again added from the burette and mixed well. The time of disappearance is recorded. This procedure is repeated 4-5 times (Table **3.8**).

3) The concentration of hydrogen peroxide is calculated at the measured time interval, after taking into account the increased volume of the reaction mixture due to the addition of hypo. The concentration of hydrogen and iodide ions changes, but this change is negligible in comparison to initial concentrations of hydrogen ions and iodide ions.

4) Initial concentration of H_2O_2 can be calculated in terms of an equivalent volume of hypo. 10 ml of concentrated sulphuric acid and 8 g of KI (dissolved in a minimum quantity of water) are added to 10 ml of H_2O_2 solution. The liberated iodine is titrated against 0.025 (M) hypo solution.

Observation

Experimental temperature = ...° C

Table 3.8. Volume of hypo solution consumed in different time interval.

Time t (sec)	**t_1**	**t_2**	**t_3**	**...**	**t_{inf}**
Vol (ml), V_t of 0.025M hypo	V_1	V_2	V_3	...	V_{inf}

Calculations

$$V_{inf} \propto [H_2O_2]_0$$

$$V_{inf} \propto V_t \,\alpha\, [H_2O_2]_t$$

A plot between V_t and t is shown in Fig. (**3.5**). The slope of tangents at a specific time represents the rate of reaction at that time.

Order of reactions can be determined by plotting differently, one a plot of $\log(V_{inf} - V_t)$ *Vs* t (Fig. **3.5**) and other $(V_{inf} - V_t)^{-1}$ *vs* t (Fig. **3.6**).

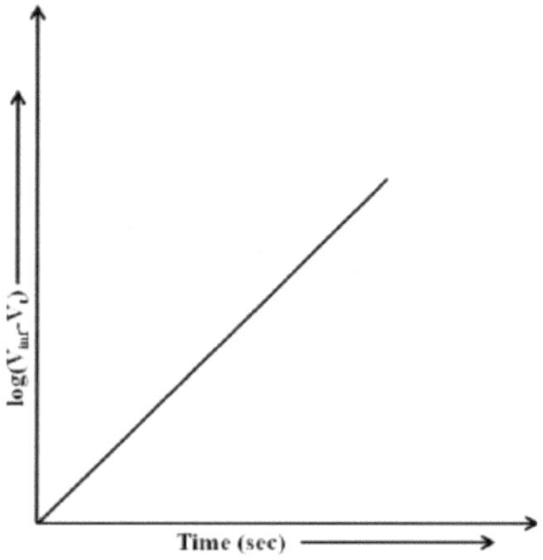

Fig. (3.5). Plot of log $(V_{inf}-V_t)$ *Vs* t.

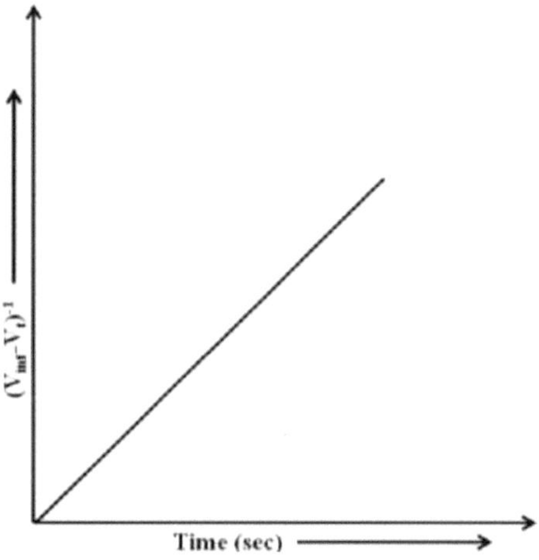

Fig. (3.6). Plot of $(V_{inf}-V_t)^{-1}$ *Vs* t.

VI. TO STUDY THE KINETICS OF BROMINATION OF PHENOL BY BROMIDE, BROMATE MIXTURE IN AN ACID MEDIUM AS A CLOCK REACTION

Chemicals and Equipment Required

0.005 (M) $KBrO_3$, 0.01 (M) KBr, 0.001 (M) phenol, 1(M) H_2SO_4, methyl orange indicator, stopwatch, 10 cm^3 pipette, distilled water, rubber tubing, beakers 100 ml, burette 50 ml.

Theory

The whole experiment involves mainly two steps. The first step involved the formation of bromine, which then reacts with phenol to form tribromophenol. Methyl orange plays the role of indicator which converts reddish-brown bromine to a colourless product.

$$5KBr+KBrO_3+3H_2SO_4 \rightleftharpoons 3H_2O+3Br_2+3K_2SO_4$$

$$3Br_2+C_6H_5OH \rightleftharpoons Br_3C_6H_2OH+3H^+Br^-$$

Thus, we focus on changing concentrations of bromide, bromate, and hydrogen ion. The initial rate of reaction in each cycle is directly proportional to $(C^x_{Br^-})_0 \left(C^y_{BrO_3^-}\right)_0 \left(C^x_{H^+}\right)_0$.

$$\frac{dc_{Br_2}}{dt}=k(C^x_{Br^-})_t \left(C^y_{BrO_3^-}\right)_t \left(C^x_{H^+}\right)_t \qquad \textbf{(3.34)}$$

Concentrations of bromide and bromate ions have been taken to remain constant in each cycle of experiments, so the initial rate can be taken as proportional to the concentration of H^+ ion.

Initial rate,

$$r_t = \left(C^z_{H^+}\right)_0$$
$$\text{or, } \ln r_t = z\ln\left(C^z_{H^+}\right)_0 + \ln(\text{constant})$$

$$\text{or, } \log r_t = z\log\left(C^z_{H^+}\right)_0 + \log(\text{constant}) \qquad \textbf{(3.35)}$$

Plot of $\log r_t$ *vs* . $\log \left(C^{z}_{H^+}\right)_t$ will be a straight line, the intercept gives the initial concentration of acid ion, and slope (z) gives the order of reaction, which is a fractional number, suggesting it to be a complex reaction with fractional order.

Procedure

Variation of Initial Rate with Concentration

1) Dilute the phenol solution with distilled water by taking 5 ml of the solution in a measuring flask.

2) This gives 0.00005 (M) phenol and is labeled as A. Also, dilute the 40 ml of 1 (M) H_2SO_4 and 2 ml of methyl orange to 100 ml in another measuring flask and label it as B. Using two 100 ml beakers, conduct the experiment. The time of the disappearance of the color will be found to be inversely proportional to KBr volume. The following two experiments (Tables **3.9** and **3.10**) are performed using two 100 ml beakers.

Experimental temperature = ……° C

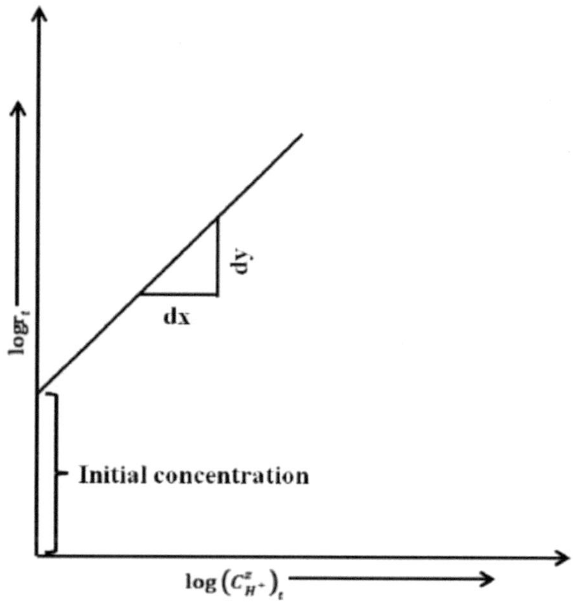

Fig. (3.7). Plot of $\log r_t$ *vs* $\log \left(C^{z}_{H^+}\right)_t$.

Table 3.9. Solution Preparation.

Beaker 1			Beaker 2			Time (Sec)
S. No.	KBr Solution (ml)	Water (ml)	$KBrO_3$ Solution (ml)	H_2SO_4 Solution (ml)	Phenol Solution (A) (ml)	
1	10	0	10	15	1	t_1
2	8	2	10	15	1	t_2
3	6	4	10	15	1	t_3
4	5	5	10	15	1	t_4
5	4	6	10	15	1	t_5

Table 3.10. Change of Initial rate with bromate ion concentration.

Beaker 1			Beaker 2			Time (Sec)
S. No.	$KBrO_3$ Solution (ml)	Water (ml)	KBr Solution (ml)	H_2SO_4 Solution (ml)	Phenol Solution (A) (ml)	
1	10	0	10	15	1	t_1
2	8	2	10	15	1	t_2
3	6	4	10	15	1	t_3
4	5	5	10	15	1	t_4
5	4	6	10	15	1	t_5

The temperatures of both solutions are kept constant. A solution from beaker 1 is added to the solution in beaker 2. Start stopwatch immediately when the beaker is emptied. The time is recorded when the solution turns colourless. It has been observed that the time is inversely proportional to the volume of potassium bromate solution consumed.

Change of Initial Rate with Acid Concentration

A sulphuric acid solution is prepared without adding methyl orange and the concentrations of KBr and $KBrO_3$ are increased significantly higher than the concentration of sulphuric acid. The concentration of phenol is kept similar as in the previous cases. For this set of experiments, 0.1 (M) sulphuric acid, 0.2 (M) $KBrO_3$, and 0.00005 (M) phenol solutions are prepared. Solution C (250 ml) is prepared by adding 12 g KBr in 5 ml of methyl orange solution. It is nearly 0.4 (M) in KBr. These solutions are arranged in two beakers as per the following Table **3.11**.

Table 3.11. Preparation of solution for monitoring iodine clock reaction.

Beaker 1			Beaker 2			Time (Sec)
S. No.	H_2SO_4 Solution (ml)	Water (ml)	$KBrO_3$ Solution (ml)	KBr Solution (ml)	Phenol Solution (A) (ml)	
1	10	0	10	15	1	t_1
2	8	2	10	15	1	t_2
3	6	4	10	15	1	t_3
4	5	5	10	15	1	t_4
5	4	6	10	15	1	t_5

The initial rate of reaction is very fast when 10 ml sulphuric acid is used. Initial rate decreases with decreasing volume of sulphuric acid. However, it is not inversely proportional to the volume of sulphuric acid taken.

Calculations

The rate for all the sets has been calculated and graph between log r_t and log $\left(C^{z}_{H^+}\right)_0$(Fig. **3.7**) has been plotted. The order of reaction has been calculated with the help of the slope of the straight line.

Precautions

Avoid direct contact of phenol with skin.

VII. TO FIND OUT ORDER OF REACTION BETWEEN POTASSIUM BROMATE AND POTASSIUM IODIDE

Chemicals and Apparatus Required

(M/10) of all potassium bromate ($KBrO_3$), potassium iodide (KI) and hydrochloric acid, and (M/100) sodium thiosulphate or hypo solution ($Na_2S_2O_3$, $5H_2O$). Thermostat, conical flask, starch solution, burette, pipette and measuring cylinder, and distilled water.

Theory

$KBrO_3$ reacts with KI in an acidic medium to give KBr along with elemental iodine. The reaction takes place as follows:

$$KBrO_3 + 6\ KI + 6HCl = 6KCl + KBr + 3H_2O + 3I_2$$

The rate of reaction can be studied by titrating the reaction mixture with hypo solution as iodine is liberated during this reaction. The titer value at any time, t, is a measure of $KBrO_3$ used for oxidation or KI oxidized. So, it gives the value of x at different time intervals.

The overall reaction is of second order. However, the molecularity is much higher. If the initial concentration of both the reactants is the same, then the rate constant can be expressed as:

$$k=\frac{1}{at}\frac{x}{(a-x)} \quad \textbf{(3.36)}$$

Where, a and x are the initial concentrations of reactant and amount of reactant reacted after time t, respectively.

Procedure

1) Take a mixture of 25 ml of (M/10) KI, 100 ml HCl of the same strength, and 100 ml distilled water in one flask, and 25 ml (M/10) $KBrO_3$ in another flask. Keep both the flasks in a thermostat till they attain a constant temperature. Now transfer $KBrO_3$ into the flask containing a mixture of KI, HCl, and distilled water. Record the time by when half the volume of $KBrO_3$ solution gets discharged. This is the starting time of the reaction.

2) Withdraw 25 ml reaction mixture and transfer it to a conical flask containing some pieces of ice to arrest the reaction. Titrate the liberated iodine against (M/100) hypo solution. Add ~1 ml of starch indicator when the colour of the titrant becomes straw yellow. Withdraw the reaction mixture after regular intervals, say 5 min, 10 min up to 40 min and titrate it as described earlier (Table **3.12**).

3) Carry a similar experiment with the following solutions: 12.5 ml (M/10) $KBrO_3$, 12.5 ml (M/10) KI, 100 ml (M/10) HCl, and 125 ml distilled water.

Observations

Experimental temperature=.... ° C

Table 3.12. Titration by Hypo Solution.

Experiment 1		**Experiment 2**	
Time (min)	**Vol. of hypo (ml)**	**Time (min)**	**Vol. of hypo (ml)**
5		5	
10		10	
15		15	

.....		.....	
40		40	

Calculations

Part I: Experiment 1

In the first experiment total volume of solution containing 25 ml (M/10) $KBrO_3$ is made to 250 ml. Hence, the normality of $KBrO_3$ in the reaction mixture changes to (M/100). 25 ml of the reaction mixture is titrated against hypo. From the normality equation,

$$M_2V_2=M_1V_1.$$

The first term stands for reaction mixture and the second term for hypo solution.

$$\frac{M}{100}\times V_2=\frac{M}{100}\times 25$$

$$V_2=25\text{ml}$$

Thus, a = b = 25 ml of (M/100) hypo ≡ 0.1 mole/lit

Part I: Experiment II

For this experiment also a=b=12.5 ml (M/100) hypo≡0.05 mole/lit. The value of the rate constant for both the experiments can be calculated by the expression: $k=\frac{1}{at}\frac{x}{(a-x)}$. The significance of the alphabets is mentioned above. The values of x/(a-x) can be expressed in terms of volumetric readings. The values of k are calculated at different time intervals and it comes out to be constant.

Plot the graph between titre values (x) *versus* t for both experiments. Draw smooth curves and find the times (t_1 and t_2) required to complete any fraction, say, 25% completion of the reaction. Calculate the order of reaction according to the equation:

$$n=1+\frac{\log\left(\frac{t_1}{t_2}\right)}{\log\left(\frac{a_2}{a_1}\right)} \quad \textbf{(3.37)}$$

Alternately, if we plot a curve between time as ordinate and 1/(a-x) as abscissa, we get a straight-line indicative of the reaction to be of the second order.

Results

The reaction between $KBrO_3$ and KI is of second-order and value of rate constant k=.... $Mol.lit^{-1}min^{-1}$

VIII. TO STUDY THE AUTO CATALYTIC REACTION BETWEEN PERMANGANATE AND OXALATE IONS CATALYSED BY MANGANESE IONS

Chemicals and Apparatus Required

0.1 (M) and 0.2 (M) oxalic acid ($H_2C_2O_4$, Ox), 0.02 (M) potassium permanganate ($KMnO_4$), 0.1 (M) sulphuric acid, 0.2 (M) manganous sulphate ($MnSO_4$), 0.1 (M) hypo, 0.1 (M) potassium iodide (KI), burette, pipette, conical flask, measuring cylinder, freshly prepared starch solution, stopwatch.

Theory

$KMnO_4$ reacts with oxalic acid in presence of mineral acid as follows:

$$2KMnO_4+5H_2C_2O_4+3H_2SO_4=K_2SO_4+2MnSO_4+10CO_2+8H_2O$$

Mn^{2+} acts as an auto-catalyst in this reaction. The progress of the reaction is followed by estimating residual MnO_4^- ion concentration from time to time by running the samples of the mixture to an excess of KI and then titrating the liberated iodine against standard hypo solution.

Procedure

1) Three reaction mixtures (consisting of a total volume of 200 ml) are prepared as follows:

a) 100 ml 0.1 (M) Ox + 30 ml 0.02 (M) $KMnO_4$ + 10 ml 0.1 (M) H_2SO_4 + No $MnSO_4$ + 60 ml H_2O.

b) 100 ml 0.1 (M) Ox + 30 ml 0.02 (M) $KMnO_4$ + 20 ml 1(M) H_2SO_4 + 20 ml 0.2(M) $MnSO_4$ + 40 ml H_2O.

c) 100 ml 0.2 (M) Ox + 15 ml 0.02 (M) $KMnO_4$ + 10 ml 1(M) H_2SO_4 + 20 ml 0.2(M) $MnSO_4$ + 55 ml H_2O.

2) Add 50 ml 0.2 (M) $KMnO_4$ to the mixture for each experiment and note down the time. Shake well and withdraw 10 ml reaction mixture into a conical flask. Again, note the time and immediately add 10 ml 0.1 (M) KI. This reacts with $KMnO_4$ and stops the reaction. The liberated iodine is estimated by titrating with 0.1 (M) hypo. A starch indicator is added as mentioned in the previous experiment.

3) Take 10 ml of the reaction mixture after every 5 min and estimate liberated iodine as described earlier. Continue the process till the titre values become very small, indicating that the liberation of iodine is very less. Plot a graph between titre values as ordinate and time (min) as abscissa.

Observation

Plot titre value of hypo against time for mixtures 1, 2, and 3. In mixtures 2 and 3, Mn^{2+} ions are present in excess. In the case of mixtures 2 and 3, titre value falls rapidly from the very beginning and autocatalysis disappears. The catalytic effect of Mn^{2+} ions can be observed from the shape of the curve for reaction mixture 1. In this case, the reaction, in the beginning, is very slow, but the rate increases rapidly due to the formation of Mn^{2+} ions.

Calculations

The time required for the half change can be determined from the time concentration graph for mixtures 2 and 3. The order of reaction with respect to $KMnO_4$ can be calculated using the equation

$$n=1+\frac{\log\left(\frac{t_1}{t_2}\right)}{\log\left(\frac{a_2}{a_1}\right)} \quad \textbf{(3.38)}$$

IXA. STUDIES ON KINETICS OF IODINATION OF ACETONE [IODINE EFFECT]

Apparatus and Chemicals Required

Burette, Pipette, Conical flask, Volumetric flask, Acetone, Potassium iodide solution, Iodine, Sulphuric acid, Sodium bicarbonate, Starch solution.

Theory

Iodine reacts with acetone to form acetonyl iodide.

$$CH_3COCH_3 \rightarrow I_2 \rightarrow CH_3COCH_2I + HI$$

The reaction is catalysed by acid. If acetone is in large excess, its concentration may be constant. The reaction is zero-order w.r.t Iodine.

$$\text{Rate} = \frac{dy}{dx} = K\,[I]^{o}$$

$$x = Kt + \text{constant} \qquad \textbf{(3.39)}$$

Where x is the concentration of I_2 consumed at any time t.

Let, V_o and V_t be the volumes of thiosulphate required corresponding to the initial concentration of iodine and the concentration of iodine at time t. Then, $x = V_o - V_t$

$$V_o - V_t = Kt + \text{constant} \qquad \textbf{(3.40)}$$

Measurement of V_o is different as the reaction starts as soon as iodine is mixed with acetone and it may be eliminated by taking the different $\Delta t_n = t_n - t_1$, such that:

$$V_1 - V_x = K\Delta t_n \quad [n = 2, 3, 4 \ldots\ldots.]$$

$$\Delta V_x = K\Delta t_n \qquad \textbf{(3.41)}$$

Where, $\Delta V_x = V_1 - V_x$ and $\Delta t_n = t_n - t_1$.V_1 and V_x are titre values after time t_1 and t_n. Since the reaction is zero-order w.r.t. iodine, the rate-determining step does not involve the interaction between the ketone and iodine molecule. So, in reality, what is being measured is the rate of enolisation of acetone.

Procedure

1. 0.1 N I_2 solution is prepared by dissolving 12.7 g of I_2 in a saturated solution of 16.6 g of KI. 50 ml of 0.1N I_2 solution is taken and diluted to 100 ml to make a 0.05 N solution.

2. 250 ml of each of N/50 $Na_2S_2O_3$, N/50 $NaHCO_3$, 100 ml of 1.0 N H_2SO_4 is prepared and also 50 ml of a starch solution is prepared.

3. The following composition of sets are prepared:

Stoppered Bottle no.	Vol. of Distilled Water (ml)	Vol. of Acetone (ml)	Vol. of 1 N H_2SO_4 (ml)	Total Volume (ml)
1	100	20	10	130
2	100	20	10	130

4. 25 ml of 0.1N I_2 is added to set 1, starting the stopwatch when the pipette is half–emptied. 10 ml of the reaction mixture is withdrawn at an interval of 3 min and run into 10 ml 0.1N $NaHCO_3$ and the time of half discharge is noted (Table **3.13**).

5. N/50 $Na_2S_2O_3$ is titrated using starch as an indicator. The readings are noted when the reaction mixture becomes colourless.

6. 25 ml of 0.1 N I_2 is added to set 2 and the same procedure is repeated (Table **3.14**).

7. ΔV_x is plotted against Δt_n for the two sets on the same graph. K_1, K_2 is determined and the ratio K_1/K_2 is found.

Calculation

Table 3.13. Titration results (For set 1).

Room Temperature = ……. ^{o}C

No. of Observations	Vol. of Reaction Mixture (ml)	Time (sec)	Δt_n(sec)	Vol. of N/50 thio Required (ml)	ΔV_x(ml)

Table 3.14. Titration results (For set 2).

No. of Observations	Vol. of Reaction Mixture (ml)	Time (sec)	Δt_n(sec)	Vol. of N/50 Thio Required (ml)	ΔV_x(ml)

From the graph,

$K_1 =$ ______________

$K_2 =$ ______________

The ratio $K_1/K_2 =$ _________

Result

The observed value of the ratio is _______

IXB. STUDIES ON KINETICS OF IODINATION OF ACETONE [ACID EFFECT]

Apparatus and Chemicals Required

Burette, Pipette, Conical flask, Volumetric flask, Acetone, Potassium iodide solution, Iodine, Sulphuric acid, Sodium bicarbonate.

Theory

Iodine reacts with acetone to form acetonyl iodide.

$$CH_3COCH_3 \rightarrow I_2 \rightarrow CH_3COCH_2I + HI$$

The reaction is catalysed by acid. If acetone is in large excess, its concentration may be constant. The reaction is zero-order w.r.t Iodine.

$$\text{Rate} = \frac{dy}{dx} = K\,[I]^{o}$$

$$x = Kt + \text{constant} \qquad \textbf{(3.39)}$$

Where x is the concentration of I_2 consumed at any time t.

Let, V_o and V_t be the volumes of thiosulphate required, corresponding to the initial concentration of iodine and the concentration of iodine at times t,

Then, $x \times V_o - V_t$

$$V_o - V_t = Kt + \text{constant} \qquad \textbf{(3.40)}$$

Measurement of V_o is different as the reaction starts as soon as iodine is mixed with acetone and it may be eliminated by taking the different $\Delta t_n = t_n - t_1$, such that:

$$V_1 - V_x = K\Delta t_n \;\; [n = 2, 3, 4........]$$

$$\Delta V_x = K\Delta t_n \qquad \textbf{(3.41)}$$

Where $\Delta V_x = V_1 - V_x$ and $\Delta t_n = t_n - t_1$. V_1 and V_x are titre values after time t_1 and t_n.

Since the reaction is of zero-order w.r.t. iodine, the rate-determining step does not involve the interaction between the ketone and iodine molecule. So, in reality, what is being measured is the rate of enolisation of acetone.

Procedure

1. 0.1N I_2 solution is prepared by dissolving 12.7g of I_2 in a saturated solution of 16.6 g of KI. 50 ml of 0.1N I_2 solution is taken and diluted to 100 ml to make a 0.05N solution.

2. 250 ml of each of N/50 $Na_2S_2O_3$, N/50 $NaHCO_3$, 100 ml of 1.0N H_2SO_4 is prepared and also 50 ml of starch solution is prepared.

3. The following composition of sets are prepared:

Stoppered Bottle no.	Vol. of Distilled Water (ml)	Vol. of Acetone (ml)	Vol. of 1 N H_2SO_4 (ml)	Vol. of 0.5N H_2SO_4 (ml)
1	100	20	10	-
2	100	20	-	10

4. 25 ml of 0.1N I_2 is added to set 1, starting the stopwatch when the pipette is half–emptied. 10 ml of the reaction mixture is withdrawn at an interval of 3 min and run into 10 ml 0.1 N $NaHCO_3$ and the time of half discharge is noted (Table **3.15**).

5. N/50 $Na_2S_2O_3$ is titrated using starch as an indicator. The readings are noted when the reaction mixture becomes colourless.

6. 25 ml of 0.1N I_2 is added to set 2 and the same procedure is repeated (Table **3.16**).

7. ΔV_x is plotted against Δt_n for the two sets on the same graph. K_1, K_2 is determined and the ratios K_1/K_2 are found.

Calculation

Table 3.15. Titration results (For set 1).

Room temperature = …... °C

No. of Observations	**Vol. of Reaction Mixture (ml)**	**Time (sec)**	**Δt_n(sec)**	**Vol. of N/50 Thio Required (ml)**	**ΔV_x(ml)**

Table 3.16. Titration results (For set 2).

No. of Observations	**Vol. of Reaction Mixture (ml)**	**Time (sec)**	**Δt_n(sec)**	**Vol. of N/50 thio Required (ml)**	**ΔV_x(ml)**

From the graph,

K_1 = ______________

K_2 = ______________

The ratio K_1/K_2 = __________

Result

The observed value of the ratio is _______

IX. C. STUDIES ON KINETICS OF IODINATION OF ACETONE [ACETONE EFFECT]

Apparatus Required

Burette, Pipette, Conical flask, Volumetric flask.

Chemicals Required

Acetone, Potassium iodide solution, Iodine, Sulphuric acid, Sodium bicarbonate, Starch solution.

Theory

Iodine reacts with acetone to form acetonyl iodide.

$$CH_3COCH_3 + I_2 \rightarrow CH_3COCH_2I + HI$$

The reaction is catalysed by acid. If acetone is in large excess, its concentration may be constant. The reaction is zero-order w.r.t Iodine.

$$\text{Rate} = \frac{dy}{dx} = K\,[I]^o$$

Integrating the above equation, we get:

$$x = Kt + \text{constant} \qquad \textbf{(3.39)}$$

Where x is the concentration of I_2 consumed at any time t.

Let, V_o and V_t be the volumes of thiosulphate required, corresponding to the initial concentration of iodine and the concentration of iodine at times t,

Then, $x \times V_o - V_t$

$$V_o - V_t = Kt + \text{constant} \qquad \textbf{(3.40)}$$

Measurement of V_o is different as the reaction starts as soon as iodine is mixed with acetone and it may be eliminated by taking the different $\Delta t_n = t_n - t_1$, such that:

$$V_1 - V_x = K\Delta t_n \quad [n = 2, 3, 4........]$$

$$\Delta V_x = K\Delta t_n \tag{3.41}$$

Where $\Delta V_x = V_1 - V_x$ and $\Delta t_n = t_n - t_1$. V_1 and V_x are titre values after time t_1 and t_n. Since the reaction is zero-order w.r.t. iodine, the rate-determining step does not involve the interaction between the ketone and iodine molecule. So, in reality, what is being measured is the rate of enolisation of acetone.

Procedure

1) 0.1N I_2 solution is prepared by dissolving 12.7g of I_2 in a saturated solution of 16.6 g of KI. 50 ml of 0.1N I_2 solution is taken and diluted to 100 ml to make a 0.05N solution.

2) 250 ml of each of N/50 $Na_2S_2O_3$, N/50 $NaHCO_3$, 100 ml of 1.0N H_2SO_4 is prepared and also 50 ml of a starch solution is prepared.

3) The following composition of sets is prepared.

Stoppered Bottle No.	Vol. of Distilled Water (ml)	Vol. of Acetone (ml)	Vol. of 1N H_2SO_4 (ml)	Total Volume (ml)
1	100	20	10	130
2	110	10	10	130

4) 25 ml of 0.1N I_2 is added to set 1, starting the stopwatch when the pipette is half–emptied. 10 ml of the reaction mixture is withdrawn at an interval of 3min and run into 10 ml 0.1N $NaHCO_3$ and the time of half discharge is noted (Table **3.17**).

5) N/50 $Na_2S_2O_3$ is titrated using starch as an indicator. The readings are noted when the reaction mixture becomes colourless (Table **3.18**).

6) 25 ml of 0.1N I_2 is added to set 2 and the same procedure is repeated (Table **3.14**).

7) ΔV_x is plotted against Δt_n for the two sets on the same graph. K_1, K_2 is determined and the ratio K_1/K_2 is found.

Calculation

Room temperature = ……. °C

Table 3.17. Titration results (For set1).

No. of Observations	Vol. of Reaction Mixture (ml)	Time (sec)	Δt_n(sec)	Vol. of N/50 thio Required (ml)	ΔV_x(ml)

Table 3.18. Titration results (For set2).

No. of Observations	Vol. of Reaction Mixture (ml)	Time (sec)	Δt_n(sec)	Vol. of N/50 thio Required (ml)	ΔV_x(ml)

From the graph,

K_1 = ______________ and K_2 = ______________ The ratio K_1/K_2 = _________

Result

The observed value of the ratio is _______

FURTHER READING

[1] Laidler, K. J. "Chemical Kinetics" 3rd Edition, Pearson Education, Ottawa, Canada, 1997.

[2] Atkins, P. W.; de Paula, J. "Atkins' Physical Chemistry" 11th Edition, Oxford University Press, 2018.

[3] Rakshit, P. C. "Physical Chemistry" 7th Edition, Sarat Book House, Calcutta, India, 2014.

[4] Palit S. R.; De, S. K. "Practical Physical Chemistry" Science Book Agency, Kolkata, 1974.

[5] Daniels, F.; Mathews, J. H.; Williams, J.W.; Bender, P.; Alberty, R. A. "Experimental physical chemistry" McGraw-Hill Book Company INC, New York, Toronto-London, 1956.

[6] Athawale, V. D.; Mathur, P. "Experimental Physical Chemistry" New Age International (P) Limited Publishers, New Delhi, 2011.

CHAPTER 4

Properties of Liquid

I. TO MEASURE THE DENSITY OF A LIQUID USING A PYKNOMETER

Chemicals and Apparatus Required

Pyknometer, given liquid, distilled water.

Theory

The density of a given liquid (d_t) at a certain temperature is given by the following formula:

$$d_t = \frac{W'D}{W} - \frac{0.012}{W(W'-W)} \quad \textbf{(4.1)}$$

W', W and D are the weight of a given volume of the liquid, the same of water and density of water at temperature t° C, respectively. A small fraction is subtracted due to the correction of buoyancy of the air, which can be neglected for ordinary laboratory work.

Procedure

1) Clean the pyknometer properly with chromic acid, wash with water thoroughly and take weight after drying. Let this weight be W_1 g.

2) Fill the pyknometer with water, using a hypodermic syringe or an aspirator. Else, attach a rubber tube at the end of part A of the pyknometer. Place the other end, B, in water and suck gently. In case a volatile or poisonous liquid is used in place of water, interpose a vapour trap between the liquid and mouth. (A drying tube filled with activated carbon is a convenient and effective trap). If any air bubble is trapped, tilt the pyknometer, so that the bubble is at the entrance to the outlet tube. Then suck more liquid into the pycnometer.

3) Ensure that the water stands at A. To remove excess water, tilt the pyknometer until a drop forms at the B end and the water level comes to A. Then wipe the drop away with a filter paper. This adjustment of the water level can be most conveniently done by slightly tilting the pyknometer toward B and withdrawing the

Charu Arora and Sumantra Bhattacharya

excess water by touching the B-end with a piece of filter paper. Add water if necessary. Dry the outer wall of the pyknometer and take the weight. Let that weight be W_2 g.

4) Replace water with the liquid under investigation. Repeat the above procedure and weigh the pyknometer with the given liquid. Let the weight be W_3 g.

Observation and Results

Temperature = ...° C

W_1= Mass of pyknometer

W_2= Mass of pyknometer + water

W_3= Mass of pyknometer + liquid

Therefore, $(W_2 - W_1) = W$= Mass of water at t° C

$(W_3 - W_1) = W'$= Mass of liquid at t° C

The density of water at t° C = D

Thus, the Density of liquid (d_t) = (W'/W) *D.

II. TO MEASURE THE DENSITY OF A LIQUID USING A DENSITY BOTTLE

Chemicals and Apparatus Required

Density bottle, given liquid, distilled water.

Theory

Same as Experiment I.

Procedure

1) Clean the density bottle properly with chromic acid, wash with water thoroughly and take weight after drying. Let this weight be W_1 g.
2) Fill the density bottle with water using a burette and note the volume. Dry the outer wall of the density bottle and take the weight. Let that weight be W_2 g.

3) Replace water with the liquid under investigation. Repeat the above procedure and weigh the density bottle with the given liquid. Let the weight be W_3 g.

Observation and Results

Same as of previous experiment (Experiment I).

SURFACE TENSION OF LIQUID

General Introduction

In liquid, the molecules have an intermolecular force of attraction. The liquid molecules can be classified into two parts. One is bulk and another is the surface of the liquid. The bulk molecules experience only attraction between liquid molecules, so the force is uniform throughout, whereas the surface molecules experience attraction forces from liquid molecules as well as vapor molecules. Since the number of vapor molecules is much less than the liquid counterpart, the attraction is unbalanced. As a result, the surface molecules of liquid behave like a stretched membrane, with a tendency to contract to a minimum area. Such property of a liquid is known as the surface tension of the liquid, which is a measure of inward pulls resulting from the unbalanced force on the surface molecules.

Definition

The force acting along the surface of a liquid at a right angle to an imaginary line of unit length drawn on the surface at a definite temperature is called the surface tension of the liquid. Generally, Surface tension is denoted by γ.

SI unit: N-m^{-1}, **CGS unit:** dyne-cm^{-1}.

Surface Energy

To increase the surface area more molecules will have to be brought to the surface. This requires some work to be done. Thus, surface energy is defined as the work done in increasing the surface area by unity, which is an alternative definition of surface tension. Thus, surface tension and surface energy have the same unit.

The surface tension of a liquid decreases with the increase in temperature. The relationship can be explained using the Ramsay-Shields-Eötvös equation. The concentration of a liquid has an effect on surface tension. The change is purely individual and depends on the nature of the solute.

Stalagmometer

A stalagmometer consists of a pipette-like glass tube with a bulb in the middle and a fine capillary with a sharp edge. At the upper end, a rubber tube is connected for sucking. A pinch-cock is arranged to control the fall of a liquid drop. The desirable drops rate should be controlled for each experiment as per requirement (generally 10-15 drops).

III. TO STUDY THE SURFACE TENSION OF A LIQUID AND DETERMINE THE SURFACE TENSION OF AN UNKNOWN LIQUID

Chemicals and Apparatus Required

Water, given liquid, stalagmometer, pyknometer, or density bottle.

Theory

The surface tension (γ_2) of the given liquid can be calculated according to the following expression:

$$\gamma_2 = \frac{n_1 d_2}{n_2 d_1} \gamma_1 \tag{4.2}$$

Letters with subscript 1 correspond to the water and 2 for the given liquid. n represents number of drops and d stands for density.

Stalagmometer

The stalagmometer is shown in Fig. (**4.1**). It contains a two-way opening and a bulb is in between. The B side of the apparatus has a capillary tube and it flattens out to give a larger surface and the surface is ground flat and polished. The capillary is sealed onto a wider tube of the wider bore on which a bulb is blown and on the stem of the tube two marks are etched, one above and another below the bulb. A small piece of rubber tube along with a pinch is fitted on the A-side of the instrument. The pinch is used to control the flow of liquid by limiting the influx of air.

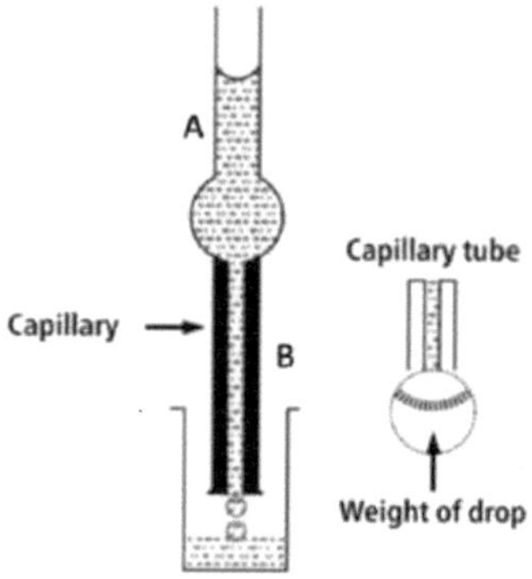

Fig. (4.1). Stlagmometer.

Procedure

1) Clean the stalagmometer thoroughly with chromic acid solution and rinse it several times with distilled water. Note down the temperature and measure the density of water and given liquid by using the aforesaid experiment.

2) The pinch clamp is opened and the dropping tip is submersed inside distilled water and sucked till the end mark of the stalagmometer A-side and the clamp is closed. Fix the stalagmometer vertically to avoid jarring it during the experiment. Adjust the pinch clamp, such that 10-16 drops come per minute, and fix the drop rate throughout the experiment. When the water meniscus touches the upper mark of the A-side, start counting drops and count till the meniscus touches the end mark of the B side. Repeat the process at least three times to have good precision.

3) Repeat the above experiment using the given liquid (Table **4.1**).

Observations and Results

Temperature = ... ^{o} C

The surface tension of distilled water at the temperature = ... γ_1 dynes/cm

The density of water = ... g/cc

The density of liquid = ... g/cc

Table 4.1. Calculation of drops of water and given liquid.

S. No.	Number of Drops				Surface Tension of Liquid (γ_2) dyne/cm
	Water (n_1)	Mean	Liquid (n_2)	Mean	
1					
2					
3					

Result

Surface tension (γ_2) of the liquid at t° C = ... dyne/cm.

Precautions

1) The stalagmometer should stand vertically.

2) The stalagmometer should not contain any greasy material.

3) Always ensure that the capillary B of the stalagmometer is not touching anything other than distilled water and experimental liquid.

4) The drop should be allowed falling off from the stalagmometer tip under its weight and should not be pushed away by the kinetic flow.

IV. TO FIND OUT SURFACE TENSION OF METHANOL, ETHANOL AND N-HEXANE AT ROOM TEMPERATURE AND CALCULATE THE ATOMIC PARACHORS OF CARBON, HYDROGEN AND OXYGEN

Chemicals and Apparatus Required

Ethanol (EtOH), methanol (MeOH), normal hexane (hexane), distilled water, stalagmometer, density bottle, beakers.

Theory

Parachor, [P] of a liquid of molar weight M and density d and surface tension γ is given by:

$$[P] = \frac{M\gamma^{1/4}}{d} \quad \textbf{(4.3)}$$

The determination of density and surface tension of all three liquids is done accurately using the above procedures. Thus, the value of the parachor of each liquid can be calculated. Atomic parachors of carbon, hydrogen and oxygen can be found as follows:

$$\text{(a) } [P_{EtOH}] - [P_{MeOH}] = [P_{CH_2}] \quad \textbf{(4.4)}$$
$$\text{(b) } [P_{hexane}] - 6[P_{CH_2}] = [P_H] \quad \textbf{(4.5)}$$
$$\text{(c) } [P_{CH_2}] - 2[P_H] = [P_C] \quad \textbf{(4.6)}$$
$$\text{(d) } [P_{MeOH}] - 4[P_H] - [P_c] = [P_O] \quad \textbf{(4.7)}$$

Procedure

The density and surface tension of all the given liquids is determined by using the above two experiments.

Note down the room temperature.

Calculations and Results

Weight of density bottle = w_1 g

Weight of density bottle + water = w_2 g

Weight of density bottle + MeOH= w_3 g

Weight of density bottle + EtOH = w_4 g

Weight of density bottle + n-hexane = w_5 g

Observation Table

Temperature = ……° C

Table 4.2. Number of Drops for Water, Methanol, Ethanol, and Hexane.

Liquid	No. of Drops	Average no. of Drops	Surface Tension of Liquid Dyne/cm
Water			

MeOH			
EtOH			
hexane			

Let molecular weights of MeOH, EtOH, and hexane be M_1, M_2, M_3, respectively.

The density of MeOH = $\frac{W_3\text{-}W_1}{W_2\text{-}W_1}$

The density of EtOH = $\frac{W_4\text{-}W_1}{W_2\text{-}W_1}$

The density of hexane = $\frac{W_3\text{-}W_1}{W_2\text{-}W_1}$

The density of distilled water can be calculated from Table **4.2** or can be measured using the above procedure.

Since we have counted the number of drops, we can calculate the surface tension of all of them. Since all the parameters of Equation (4.1) are known, atomic parachors of C, H, and O can thus be calculated using Equation (4.4)-(4.7).

Results

Report parachor values for water, methanol, ethanol and hexane (Table **4.3**). Calculate atomic parachor using Equation (4.3) and report in Table **4.4**.

Table 4.3. Parachor Values for Water, Methanol, Ethanol, and Hexane.

Compound	**Parachor Value**
EtOH	
MeOH	
n-hexane	

Table 4.4. Atomic Parachor values.

Compound	**Parachor Value**
H	
C	
O	

V. TO DETERMINE THE CRITICAL MICELLE CONCENTRATION (CMC) OF SOAP

Chemical and Apparatus Required

Distilled water, soap, stalagmometer.

Theory

Soap consists of a hydrophobic anionic part which is mainly a long chain compound and a hydrophilic cationic part. When soap is added to water, the hydrophilic cationic parts attract the anionic parts of dissolved salts in the water and lower the surface tension. If the solution is not very dilute, the surface tension of the mixture varies directly with the log of the concentration of the added substance. However, in the case of soap and other substances undergoing association, the curve shows a break, which corresponds to critical micelle concentration.

Procedure

1) Prepare a 0.2 (M) solution of potassium laurate (stock solution) (MW 238.4) in distilled water.

2) Prepare 0.1, 0.05, 0.025, 0.02, 0.01, 0.001 (M) solution of potassium laurate, made in different volumetric flask of the same volume (Table **4.5**).

3) Determine the density and surface tension of all these solutions using the method discussed in the earlier experiment (To determine the surface tension of a given liquid).

Observation

Temperature = ……° C

Table 4.5. Variation of Drops with Variation in Concentration of Micelle.

Concentration	$\log_{10}$(Concentration)	No. of Drops	Density (g/cc)	Surface Tension (dynes/cm)
0	-	-	-	-
0.2 (M)	-	-	-	-
0.1 (M)	-	-	-	-
0.05 (M)	-	-	-	-
0.025 (M)	-	-	-	-
0.02 (M)	-	-	-	-
0.01 (M)	-	-	-	-
0.001 (M)	-	-	-	-

Calculation

Plot a curve between surface tension as ordinate and $\log_{10}$(Concentration) of potassium laurate in abscissa. A break in the curve so obtained gives the critical micelle concentration (CMC) of the given soap.

Result

The CMC of soap = ……..

VI. TO COMPARE CLEANSING POWERS OF GIVEN SAMPLE OF DETERGENTS

Chemicals and Apparatus Required

Distilled water, supplied detergents, beakers, density bottle/pyknometer, stalagmometer.

Theory

Several salts are present in a dissolved or suspended state in potable water. On the other hand, a soap/detergent contains two parts; one is hydrophilic cationic part and the other is hydrophobic anionic part. The cationic part is dissolved into water and upon reaction with the dissolved salts, it forms insoluble salts and precipitated; this process reduces the surface tension of water. The more the ability to reduce the surface tension of water, the more will be the cleansing power of the given detergent. Thus, if we dissolve an equal amount of two detergents individually in a fixed volume of water (premeasured surface tension), after filtering out the water and measuring the surface tension of water, the sample with less surface tension will have more cleansing power. Since the solutions will be very dilute, we can assume the density of the solution is equal to the density of water. Thus,

$$\frac{\gamma_1}{\gamma_2} = \frac{d_1 n_2}{d_2 n_1} = \frac{n_2}{n_1} \tag{4.7}$$

Here, γ_i, d_i, and n_i are the surface tension, density, and number of drops falling from the stalagmometer.

Procedure

1) Measure the density and surface tension of the water sample very accurately.

2) Two detergents with the same mass are taken individually and they are marked as 1 and 2.

3) Take detergent no.1 and prepare 0.01%, 0.02%, 0.03%, 0.04% and 0.05% (w/v) solutions by adding it to pre-heated, pre-measured volume of water in beakers. Solutions are then made by gentle stirring, using a glass rod, and then allow it to cool down to room temperature. The suspension is allowed to stand for a while so that the insoluble impurities got settled down. Filter out the solutions. Since concentrations of solutions are very low, we can assume that the density of solutions will not deviate much from that of water. Measure the surface tension of the solution using a stalagmometer (Described earlier). Before filling up the stalagmometer, ensure that the solution is pure, *i.e.*, free from any suspended particles otherwise, these suspended particles will choke the capillary tube.

4) Perform a similar experiment using detergent no. 2 (Table **4.6**).

Observation and Results

Temperature = ...° C.

Table 4.6. Variation of Drops for Two Detergents with Change in Concentration.

Concentration (%)	No. of Drops for Solution of Detergent 1 (n_1)	No. of Drops for Solution of Detergent 2 (n_2)	$\frac{\gamma_1}{\gamma_2} = \frac{n_2}{n_1}$
0.01	-	-	-
0.02	-	-	-
0.03	-	-	-
0.04	-	-	-
0.05	-	-	-

Results

If (γ_1/ γ_2) > 1, then, the cleansing power of detergent 1 is poorer than the detergent 2. Else reverse.

VII. DETERMINATION OF SURFACE EXCESS OF AMYL ALCOHOL

Chemicals and Apparatus Required

Amyl alcohol, distilled water, stalagmometer, 50 ml volumetric flask (8-10 pieces), 250 ml volumetric flask, weighing bottle.

Theory

The bulk concentration of a solute in a solution is different from the surface concentration. This difference in concentration was thermodynamically explained by Gibbs in an equation, which is famously known as the adsorption isotherm equation (Gibbs' adsorption isotherm). The expression of the Gibbs adsorption isotherm is:

$$\Gamma=-\frac{c}{RT}\left(\frac{d\gamma}{dc}\right) \quad \textbf{(4.8)}$$

Here c, γ, Γ and R are the concentration and the surface tension of the solution at temperature T K, surface excess and universal molar gas constant, respectively. The excess concentration in terms of the number of moles of solute present per unit area for a given volume of liquid on the surface over the bulk is known as the surface excess of the solute-solvent system.

If for a volume v of the solution having area A, the number of moles of solute present on the surface is n_2 and that in the bulk be n_1, then the expression for the surface area will be:

$$\Gamma=\frac{n_2-n_1}{A} \quad \textbf{(4.9)}$$

The derivative of γ with respect to the concentration may have the (+)-ve or (-)-ve sign. It may be even zero. This parameter depends on solute-solvent. Substances like soap, detergents, weak electrolytes, *etc*. which decrease the surface tension of water with an increment of concentration have a negative value of the derivative. Hence, Γ is positive, *i.e*, a positive value of surface excess will be obtained. In other words, those substances are said to have positive adsorption. Amyl alcohol, acetic acid, butyric acids, *etc*. have various surface-active agents with positive surface excess.

Equation (4.9) can be re-expressed as:

$$\Gamma = -\frac{1}{2.303RT}\left(\frac{d\gamma}{d(\log_{10} c)}\right) \quad \textbf{(4.10)}$$

Measuring the surface tensions of a number of solutions of amyl alcohol with different concentrations, a plot of γ *Vs*. ($\log_{10}$ c) is drawn with slope $\frac{d\gamma}{d(\log_{10} c)}$

Since concentration is a fraction, thus, it is advised to take -($\log_{10}$ c) instead ($\log_{10}$ c). A plot of γ *Vs*. -($\log_{10}$ c) will be like Fig. **(4.2)**.

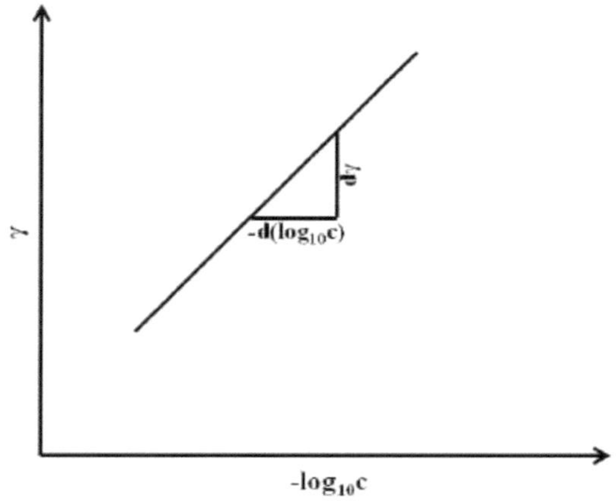

Fig. (4.2). Variation of surface tension with respect to $\log_{10}$c.

Thus, using Equation (4.10), the surface excess can be expressed as $\Gamma = \frac{1}{2.303RT}$ slope.

The surface tension is measured by using the drop-weight method.

Procedure

1) Calculate the molarity of the original amyl alcohol solution from molar weight and density by following the bottle label. Prepare 250 ml of 0.1 M amyl alcohol solution.

2) Prepare the following sets of solutions (Table **4.7**).

Table 4.7. Preparation of different concentrations of amyl alcohol solution.

No. of Volumetric Flask	Volume of 0.1 (M) Amyl Alcohol (ml)	Water Added (ml)	Final Volume After Make-up with Water (ml)	Concentration of Solution (M)
1	5	45	50	0.01
2	10	40	50	0.02
3	15	35	50	0.03
4	20	30	50	0.04
5	25	25	50	0.05
6	30	20	50	0.06
7	35	15	50	0.07
8	40	10	50	0.08

2) Note down the temperature.

3) Calculate the density of all the solutions and water (Table **4.8**).

4) Calculate the specific gravity of all the above solutions using the following relation:

$$\text{Specific gravity } (s_l) = \frac{\rho_l}{\rho_w} = \frac{\text{the density of the definite volume of solution}}{\text{the density of the same volume of water}}$$

5) Using a stalagmometer, measure accurately the surface tension of all the solutions, pure water, and amyl alcohol.

Table 4.8. Density and specific gravity of different solutions.

Concentration of Solution (M)	Weight of Empty Bottle (g)	Weight of Solution + Empty Bottle (g)	Specific Gravity= (Density of Definite Volume of Solution)/(Density of Same Volume of Water)
0.00 (Pure water)	W_0	w	1
0.01	W_0	W_1	$(W_1 - W_0)/(W-W_0)$
0.02	W_0	W_2	$(W_2 - W_0)/(W-W_0)$
0.03	W_0	W_3	$(W_3 - W_0)/(W-W_0)$
0.04	W_0	W_4	$(W_4 - W_0)/(W-W_0)$
0.05	W_0	W_5	$(W_5 - W_0)/(W-W_0)$
0.06	W_0	W_6	$(W_6 - W_0)/(W-W_0)$
0.07	W_0	W_7	$(W_7 - W_0)/(W-W_0)$
0.08	W_0	W_8	$(W_8 - W_0)/(W-W_0)$

6) Count the number of drops and calculate γ for water and all the solutions by the usual method.

Observation and Results

The surface tension of any liquid can be measured using the following relation: $\gamma_l = \frac{n_w}{n_l} s_l \gamma_l$.

Where γ_l and γ_w are surface tension of liquid and water, n_l and n_w are number of drops coming out from stalagmometer, s_l is the specific gravity of the liquid. The following table4.9 has been used to tabulate the surface tension of the liquids:

Temperature = ...° C

Table 4.9. Concentration and surface tension of the liquid.

Volumetric Flask no.	Concentration of Solution (M) c	$-\log_{10} c$	γ of the Solution (Dyne/cm)
1	0.01	2	
2	0.02	1.7	
3	0.03	1.52	
4	0.04	1.4	
5	0.05	1.3	
6	0.06	1.22	
7	0.07	1.15	
8	0.08	1.1	
9	0.1	1.0	

From the plot of γ *Vs*. $-\log_{10} c$, the slope is obtained. Hence, surface excess of amyl alcohol can be measured by using $\Gamma = \frac{1}{2.303RT} \times$slope moles/cm^2. Calculate and report an experimental error.

Literature Value

5×10^{-10} moles/cm^2.

VISCOSITY OF LIQUID

General Introduction

When a fluid (gas or liquid) flows through a tube, the liquid is assumed to be made up of a number of concentric cylindrical layers, separated by infinitesimally small distance dr apart. Due to adhesion force, the layer in contact with either of the wall remains stationary. On the other hand, the velocity of the subsequent layers approaching the center of the tube increases with velocity gradient $\frac{du}{dr}$, the central layer has the highest velocity and slowly decreases towards the wall. For the type of motion, where a layer does not perturb another layer, then the motion is known as laminar motion, else turbulent motion.

The difference in velocities may be attributed to some kind of backward force acting on a layer in motion. A relatively faster layer tries to accelerate the relatively slower layer adjacent to it and a slower layer tries to retard the relatively faster layer in contract with it. The resistance to relative motion between two layers of a liquid is known as viscosity. For stagnant fluid (liquid) viscosity plays no role. It is only valid when the fluid is in motion.

If F is the applied force for a laminar flow, the retarding force or the viscous force will be –F, which is proportional to the area of contact (A) between two moving layers and the velocity gradient. Therefore,

$$F = \propto A\frac{du}{dr}$$

$$\text{or } F = \eta A\frac{du}{dr} \quad \textbf{(4.11)}$$

Here η is the coefficient of viscosity and it is different for different liquids

Definition

Force required in maintaining unit velocity gradient between two layers of unit area of contact.

From the above equation we can define unit (CGS) of η as follows:

$$\eta = \frac{F}{A}\frac{du}{dr} = \frac{dyne}{cm^2}\frac{1}{\frac{cm/s}{cm}} = \frac{g.cm/s^2}{cm^2}s = gcm^{-1}s^{-1} = poise$$

Poiseuille Equation

If V be the volume of a liquid flowing in time t through a capillary tube of length l, radius r due to the difference of pressures at two ends, then according to Poiseuille,

$$\eta = \frac{\pi r^4 l(P_1 - P_2)}{8lV} = \frac{\pi r^4 l \Delta P}{8lV} \quad \textbf{(4.12)}$$

Where ΔP is the difference between two sides of the tube

The viscosity of liquid will increase with the increase in density. It is observed that they are directly proportional.

The viscosity of a gas increases with the increase of temperature but decreases for liquids. The viscosity of liquid and temperature follows the following relationship:

$$\ln\eta = \frac{E}{RT} + \ln A \quad \textbf{(4.13)}$$

Where A is a constant, E is the energy required by one mole of liquid to overcome the retarding force. It is also named as activation energy. Thus, the plot of T^{-1} *Vs.* ln η will be a straight line and cross the y axis at ln A.

Viscometer

The coefficient of viscosity (η) of liquid is generally measured using Ostwald viscometer. Using Poiseuille's equation, direct measurement of viscosity is difficult. Alternatively, the instrument provides a simple means to determine the relative viscosity of a liquid or a solution with reference to another liquid with known η.

The Ostwald viscometer (Fig. **4.3**) is made of glass and has two limbs joined to give it a U-like shape. One of the limbs comprises a small bulb, C, below which is an 8-10 cm long capillary, D. There are two definite marks (A, B) above and below C. On the other limb, there is a bigger bulb (E) with a wider mouth through which the experimental solution is poured down. On the mouth of the other limb, an arrangement of rubber tubing is kept to suck the solution. During the experiment, the instrument is clamped vertically and time is recorded against the passage of solution from A to B.

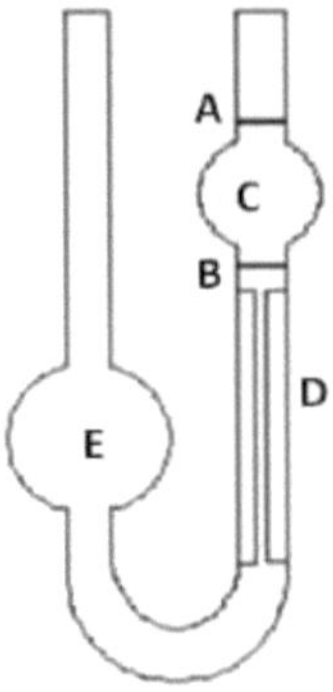

Fig. (4.3). Ostwald viscometer.

A definite amount of the experimental solution is pipetted out and poured in E. The solution is sucked through a rubber tube and allowed to fall. When the surface of the solution reaches A, the stopwatch is started on and stopped when it reaches B. Denote the time interval.

It is advisable to take the proper amount of experimental solution. An excess amount either takes much time or may not reach B at all. This adjustment is to be done by the experimentalist and is dependent on the particular viscometer. It is advised to take 15 ml solution. There are various types of viscometers. We dropped discussion on them.

VIII. TO STUDY THE COEFFICIENT OF VISCOSITY OF A LIQUID AND DETERMINE THE VISCOSITY OF AN UNKNOWN LIQUID AND VERIFICATION OF CHANGE OF VISCOSITY WITH COMPOSITION

Chemical and Apparatus

Distilled water, given solution, unknown solution, Ostwald viscometer, and pipette.

Theory

When a liquid is allowed to flow in a capillary tube in a laminar flow motion, the relative motion of the adjacent layers tries to retard due to an inherent property of the liquid, called viscosity. This is some kind of internal friction within the liquid. From Newton's law of flow of liquid, the force (F) required to maintain a velocity gradient $(\frac{du}{dr})$ between two adjacent layers is related as:

$$F = \eta A \frac{du}{dr} \quad \textbf{(4.14)}$$

Where, A is the area of contact and η is the coefficient of viscosity, which is nothing but the force required to maintain unit velocity gradient between two consecutive layers with a unit area of contact. Using the Poieseuille equation, η is determined by the following equation:

$$\eta = \frac{\pi r^4 lP}{8lV} \quad \textbf{(4.15)}$$

Here, V is the volume of a liquid flowing in unit time t through a capillary tube of length l and radius r and P is the applied average pressure. Using Ostwald viscometer, η can be determined by comparison with another liquid of known viscosity. By following the procedure:

The same volumes of two liquids are allowed to flow through the same length in a particular viscometer. Let the coefficient of viscosity of two liquids be η_1 and η_2 and time taken for flow is t_1 and t_2, respectively.

$$\text{Thus, } \eta_1 = \frac{\pi r^4 t_1 P_1}{8lV}, \quad \eta_2 = \frac{\pi r^4 t_2 P_2}{8lV}, \quad \frac{\eta_1}{\eta_2} = \frac{t_1 P_1}{t_2 P_2} \quad \textbf{(4.16)}$$

If d_1 and d_2 are the densities of liquid 1 and 2, respectively, then P = hdg. Since volume is the same, the difference in the heights of two levels in two arms of viscometer will be the same, *i.e.*, $P_1 = hd_1g$ and $P_2 = hd_2g$

Thus,

$$\frac{\eta_1}{\eta_2} = \frac{t_1 d_1 g}{t_2 d_2 g} = \frac{t_1 d_1}{t_2 d_2}$$

$$\text{Or, } \eta_1 = \eta_2 \frac{t_1 d_1}{t_2 d_2} \quad \textbf{(4.17)}$$

Here η_1 and η_1 are the viscosity coefficients of liquid 1 and 2, respectively. If η_2 is known, densities and time of flow are measured exactly, we can calculate η_1. If liquid 2 is distilled water (w) and liquid 1 is given liquid (l), if the specific gravity of the liquid is s, then, the working equation is:

$$\eta_l = \eta_w s \frac{t_w}{t_l} \; s = \frac{d_w}{d_l} \quad \textbf{(4.18)}$$

Procedure

1) Clean the viscometer thoroughly.

2) Prepare solutions of different concentrations (5%, 10%, 15%, 20%) made up of the same solute and solvent of given unknown liquid.

3) Note down the temperature.

4) Determine the specific gravity of all the solutions accurately as follows:

5) Weigh a perfectly clean, dried weighing bottle (w_0). Take 2 ml of distilled water in the bottle with the help of a 2 ml pipette and weigh (w). Add 2 ml of solution with different concentrations, the same 2 ml pipette, and weigh (w_i) (Table **4.10**).

Specific gravity can be calculated as:

$$S = \frac{\text{Weight of 2 ml solution}}{\text{Weight of 2 ml water}} = \frac{w_i - w_0}{w - w_0} \qquad \textbf{(4.19)}$$

Clamp the viscometer vertically with burette-stand (clamp the wider arm).

Using a pipette, pour down 15 ml distilled water in the wider arm. Suck the solution through a rubber tube and check whether the solution smoothly comes down and crosses the marking zone (AB). Adjust the exact volume.

After sucking water allow it to fall. Start the stopwatch when the meniscus touches the upper mark of the capillary arm and stop the watch as it just touches the lower mark. Note down the time (Table **4.11**). Repeat twice to note a good precision of water falling time (t_w).

Discard water from the viscometer, rinse the water with a solution of known concentration (say 5 %), and follow the above procedure. Calculate t_l.

Repeat the same with other solutions including the unknown and determine t_l.

Observations and Results

Temperature = ……° C

Table 4.10. Calculation of Specific Gravity.

Concentration of Solution (%) (w/w)	Weight of Empty Bottle (g)	Weight of Solution +Empty Bottle (g)	Specific Gravity=(Density of Definite Volume of Solution)/(Density of Same volume of water)
0.00 (Pure water)	W_0	w	1
5	W_0	W_1	$(W_1 - W_0)/(W-W_0)$
10	W_0	W_2	$(W_2 - W_0)/(W-W_0)$
15	W_0	W_3	$(W_3 - W_0)/(W-W_0)$
20	W_0	W_4	$(W_4 - W_0)/(W-W_0)$
Unknown	W_0	W_5	$(W_5 - W_0)/(W-W_0)$

Table 4.11. Table for Time Flow.

S. No.	Liquid	Time of Flow (Sec)	Most Precise Time (Sec)	η
1	Water			
2				
3				
1	5 % solution			
2				
3				
1	10 % solution			
2				

3				
1	15 % solution			
2				
3				
1	20 % solution			
2				
3				
1	Unknown solution			
2				
3				

Results

At .... ° C, the coefficient of viscosity of the unknown liquid is .... Poise

IX. TO FIND THE CONCENTRATION OF GIVEN MIXTURE CONSISTING OF TWO LIQUIDS A AND B BY VISCOSITY MEASUREMENTS

Chemicals and Apparatus

Distilled water, two liquids A and B, Ostwald viscometer, pipette.

Theory

When a liquid is allowed to flow in a capillary tube in a laminar flow motion, the relative motion of the adjacent layers tries to retard due to an inherent property of the liquid, called viscosity. This is some kind of internal friction within the liquid. From Newton's law of flow of liquid, the force (F) required to maintain a velocity gradient ($\frac{du}{dr}$) between two adjacent layers, is related as:

$$F = \eta A \frac{du}{dr} \tag{4.20}$$

Where, A is the area of contact and η is the coefficient of viscosity, which is nothing but the force required to maintain unit velocity gradient between two consecutive layers with a united area of contact. Using Poieseuille equation η is determined by the following equation:

$$\eta = \frac{\pi r^4 lP}{8lV} \tag{4.21}$$

Here, V is the rate of flow of the volume of a liquid through a capillary tube of length l and radius r and P is the applied average pressure. Using Ostwald viscometer η can be determined by comparison with another liquid of known viscosity by following the procedure:

The same volumes of two liquids are allowed to flow through the same length in a particular viscometer. Let the coefficient of viscosity of two liquids be η_1 and η_2 and time taken for flow is t_1 and t_2, respectively

$$\text{Thus,} \quad \eta_1 = \frac{\pi r^4 t_1 P_1}{8lV} \qquad \eta_2 = \frac{\pi r^4 t_2 P_2}{8lV} \qquad \frac{\eta_1}{\eta_2} = \frac{t_1 P_1}{t_2 P_2} \tag{4.22}$$

If d_1 and d_2 are the densities of liquid 1 and 2, respectively, then P = hdg. Since volume is the same, the difference in heights of two levels in two arms of the viscometer will be the same

$$P_1 = hd_1 g \text{ and } P_2 = hd_2 g \tag{4.23}$$

Thus,

$$\frac{\eta_1}{\eta_2} = \frac{t_1 hd_1 g}{t_2 hd_2 g} = \frac{t_1 d_1}{t_2 d_2} \qquad \eta_1 = \eta_2 \frac{t_1 d_1}{t_2 d_2} \tag{4.24}$$

Here η_1 and η_1 are the viscosity coefficients of liquid 1 and 2, respectively. If η_2 is known, densities and time of a flow are measured exactly, we can calculate η_1. If liquid 2 is distilled water (w) and liquid 1 is given liquid (l), and if the specific gravity of the liquid is s, then, the working equation is:

$$\eta_l = \eta_w s \frac{t_w}{t_l} \qquad s = \frac{d_w}{d_l}$$

By plotting the viscosity coefficient of solutions against their concentrations, we get a curve from which the concentration of the unknown solution is determined. Curves of various forms are obtained and usually, viscosity curves are sagged, *i.e.*, fall below the straight line connecting the viscosity of their components.

Procedure

1) A number of solutions are prepared by mixing two liquids, *viz.* A and B in different proportions. The solutions are made up with 90%, 80%, 70%, 60%, 50%, 40%, 30%, 20%, and 10% of A by volume.

2) Density/specific gravity of each solution is measured using the above experiment (Table **4. 10**).

3) An Ostwald viscometer is standardized using the earlier mentioned steps. In the same manner, the time of flow of all the solutions is recorded (Table **4.12**).

Observations and Results

Temperature = ...° C

Table 4.10. Calculation of specific gravity: Consult Table **4.10** of the earlier experiment for both the liquids.

Table 4.12. Calculation of time of flow.

Percentage of Component		Time of Flow (Sec.)
A (%)	B (%)	-
90	10	-
80	20	-
70	30	-
60	40	-
50	50	-
40	60	-
30	70	-

20	80	-
10	90	-
Unknown	*****	-

Calculations

A curve is plotted between concentrations of one component (say A) and time of flow in second. We see that a straight line is obtained (Fig. **4.4**). The composition of the unknown solution is calculated by plotting the time of flow on a straight line. Draw a perpendicular on the x-axis (concentration) from which the composition of the unknown solution can be measured directly.

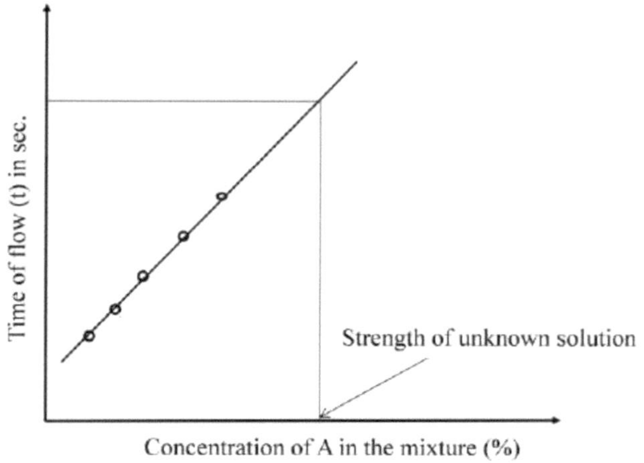

Fig. (4.4). Plot of concentration of A in the mixture in terms of wt. percentage *Vs*. Time in sec.

Result

The concentration of A is =.... %

X. TO CALCULATE MOLECULAR WEIGHT OF A HIGH POLYMER BY MEANS OF VISCOSITY MEASUREMENTS

Chemicals and Equipment

Polymer (polystyrene, preferably dried), toluene, Ostwald viscometer.

Theory

Polymers are giant molecules with a molecular weight ranging from several thousands to millions. They are soluble in suitable organic solvents and produce highly viscous solutions, even at low concentrations. Measuring the viscosities of the solutions of polymers at low concentrations, the average molecular weights of several polymers may be determined.

The molar mass M of a polymer is empirically related to its intrinsic viscosity, [η], by modified Staudinger equation or Mark-Houwink equation

$$[\eta]=KM^{\alpha} \quad \textbf{(4.25)}$$

Where K is a constant and is fixed for a polymer and also depends on the solvent and temperature. Another constant α is a function of the shape and geometry of the polymer molecule. Intrinsic viscosity is defined as:

$$[\eta] = \lim_{c\to 0}\left(\frac{\eta_{sp}}{c}\right) \quad \textbf{(4.26)}$$

c is the concentration of the polymer solution, expressed as a number of grams of polymer in 100 ml of the solution. The specific viscosity η_{sp} is defined as:

$$\eta_{sp} = \frac{\eta_{solution}}{\eta_{solvent}} - 1 \quad \textbf{(4.27)}$$

The fundamental equation for determining the relative viscosity of a solution is:

$$\frac{\eta_{solution}}{\eta_{solvent}} = \frac{\eta}{\eta_0} = \frac{dt}{d_0 t_0} \quad \textbf{(4.28)}$$

d and d_0 are the density of the solution and the solvent, respectively, and t and t_0 are the time of flow for the same. A polymer with low concentration d = d_0. So, Equation (4.28) becomes:

$$\frac{\eta}{\eta_0} = \frac{t}{t_0} \quad \textbf{(4.29)}$$

Thus, from Equation (4.27) and Equation (4.29):

$$\eta_{sp} = \frac{t}{t_0} - 1 \quad \textbf{(4.30)}$$

Thus, the plot of η_{sp} *Vs*. c, when extrapolated to zero concentration, gives [η]

Knowing K and α for a particular polymer-solvent combination, the molecular weight of the polymer can be calculated from Equation (4.25).

Sometimes [η] is determined from the relationship:

$$[\eta] = \lim_{c \to 0} \left(\frac{1}{c} \ln \left(\frac{\eta}{\eta_0} \right) \right) \quad \textbf{(4.31)}$$

Which results from the expansion of the logarithmic function $\ln\left(\frac{\eta}{\eta_0}\right)$ and neglecting the higher terms as $c \to 0$. In that case, [η] will be the intercept of the graph of $\left(\frac{1}{c}\ln\left(\frac{\eta}{\eta_0}\right)\right)$ *Vs*. c. To have a good agreement of result, it is advisable to calculate [η] from both the graphs.

Procedure

1) Prepare a 2% solution of polystyrene by accurately weighing ~2 g of dry polymer, allowing it to dissolve in a small volume of toluene and make up the volume up to the mark, *i.e.*, 100 ml.

2) Prepare the following solutions (Table **4.13**) from the 2% solution.

Table 4.13. Preparation of solutions.

Bottle no.	Vol. of 2% of Polystyrene (ml)	Volume of Toluene Added	Total Volume of Solution (ml)	Concentration of the Solution (%) (Calculate from x)
1	5.0	20.0	25.0	
2	10.0	15.0	25.0	
3	15.0	10.0	25.0	
4	19.0	6.0	25.0	
5	25.0	0.0	25.0	x

To get a better result, perform the following steps by keeping the viscometer in a thermostat, after filling up with the solutions for 10-15 min.

3) Clean and dry the viscometer thoroughly. Fill the viscometer with toluene of a particular volume and note the flow time (t_0). Repeat thrice to have a reading of good precision.

4) Rinse the viscometer with polystyrene solution of the lowest concentration. Fill up the viscometer with the same volume of the solution as above and note the flow time (t).

5) Repeat the last step with the remaining solutions with ascending concentrations

Calculate η_{sp} using Equation (4.30) and the [η] using the following Table **4.14**.

Table 4.14. Flow of time and calculation of $\eta_{sp.}$ Flow of toluene, t_0 = ……sec

Bottle no.	Conc. of the soln. c (g/100 ml) (%)	Flow Time (sec)	Mean Time Flow (sec)	$\eta_{sp} = t/t_0 - 1$	η_{sp}/c (100 ml/g)
1	-	-	-	-	-
2	-	-	-	-	-
3	-	-	-	-	-
4	-	-	-	-	-
5	-	-	-	-	-

Observation Table

Temperature = ……° C

Table 4.15. Preparation of 100 ml standard ~2% solution of polystyrene in toluene.

1st Weight of Weighing Bottle + Polystyrene w_1(g)	2nd Weight of Empty Weighing Bottle Polystyrene w_2(g)	Weight Taken (w_1-w_2) g	Conc. of the Solution After Dissolving in Toluene (%)
-	-	-	-

Results

The plot of c *Vs*. η_{sp}/c will be as Fig **(4.5)**.

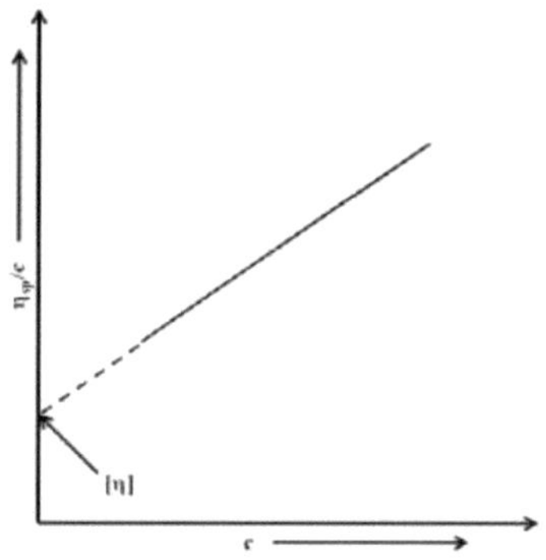

Fig. (4.5). Plot of c *vs.* η_{sp}/c.

[η] = Intercept on y-axis = …… ($100\ cm^3g^{-1}$)

Thus, from the equation $[\eta] = KM^{\alpha}$, $M = \left(\frac{[\eta]}{K}\right)^{1/\alpha}$

Put $K = 3.7*10^{-4}\ cm^3g^{-1}$ and $\alpha = 0.62$ (both at 25 ° C)

Molar mass (M) of given polymer =…. [Adjust K by dividing 100 to match the unit of η]

Note: K and α for benzene-polystyrene = $9.52*10^{-3}$ and 0.744, respectively.

XI. TO FIND OUT TEMPERATURE COEFFICIENT FOR A GIVEN LIQUID, TO DETERMINE INFLUENCE OF TEMPERATURE ON VISCOSITY

Chemicals and Apparatus

Distilled water, given liquid, Ostwald's viscometer, density bottle, thermostat, stopwatch, thermometer, glasswares.

Theory

Viscosity of liquid decreases with the increase of temperature (*cf.* The same increases for gas). The coefficient of temperature $\left(\frac{d\eta}{dt}\right)$ can be obtained by determining the absolute viscosity of the liquid compared with that of water at regular intervals of 5° C, ranging from room temperature to an elevated temperature, say 60° C. The values of viscosity are plotted against temperature and the temperature coefficient can then be calculated from the graph with an interval of 5° C.

Procedure

1) Record the room temperature.

2) Measure the density of water and the supplied liquid accurately (Procedure described earlier).

Clean and dry the viscometer thoroughly.

3) Determine the time of flow of water at room temperature using the viscometer.

4) Perform the same temperature range of (room temperature + 5)° C to a certain temperature (60° C) with an interval of the temperature of 5° C. To perform the step, clamp the viscometer vertically in a thermostat in such a manner that no obstruction takes place.

5) Perform steps 3 and 4 for the given liquid and record the time of flow as given in Table **4.16**.

Observation

Table 4.16. Record of time of flow with respect to change in temperature.

Temperature ° C	Time of Flow of Liquid (sec)	Time of Flow of Water (sec)
T	…..	…..
(t+5)	…..	…..
…..	…..	…..
60	…..	…..

Calculations

Density of liquid at t° C = d_l, Density of water at t° C=d_w.

Viscosity of liquid [η_l] = $d_l t_l \eta_w / d_w t_w$.

The viscosity of the liquid can be determined at different temperatures, using the same expression, and the viscosity of water can be taken from reference data.

Plot temperature *Vs.* absolute viscosity. Calculate the value of temperature coefficient between any range of temperature as follows:

$$\text{Temperature coefficient} = \frac{\text{viscosity at } t_1{}^{o}\,C - \text{viscosity at } t_2{}^{o}\,C}{t_1 - t_2}$$

Results

Temperature coefficient for the given liquid = ...

DISTRIBUTION PROPERTIES

General Introduction

Distribution law is known as the chemical potential of a component present in different phases of a system at equilibrium is the same in every phase. If a substance be present in two liquid phases, denoted by suffix I and II at equilibrium, its chemical potential will be the same, *i.e.*, $\mu_I = \mu_{II}$

Since $\mu = \mu^0 + RT\ln a$, where, μ, μ^0, are the chemical potentials of a component in the solution phase and the same in the pure state respectively. The universal molar gas constant is R, T is the absolute temperature, and the last term is the activity.

Thus, the above equation can be represented as:

$$\mu_I{}^0 + RT\ln a_I = \mu_{II}{}^0 + RT\ln a_{II}$$

$$\text{or, } RT\ln(a_I/a_{II}) = \mu_{II}{}^0 - \mu_I{}^0$$

$$\text{or, } \ln(a_I/a_{II}) = \frac{\mu_{II}{}^0 - \mu_I{}^0}{RT}$$

$$\text{or, } a_I/a_{II} = \exp\left(\frac{\mu_{II}{}^0 - \mu_I{}^0}{RT}\right) \qquad \textbf{(4.32)}$$

Thus, for an isothermal process (T is constant) the ratio will be constant (K). Thus, we can express the equation as: $a_I/a_{II} = K$. Thus, if the activity of the components in one phase is fixed, its activity in another phase at equilibrium is also fixed automatically. For a dilute solution, the activity is equivalent to concentration. Thus, $c_I \equiv a_I$ and $c_{II} \equiv a_{II}$. The ratio of the concentration of I and II is known as the distribution coefficient or partition coefficient (K_D) of the solute between the two immiscible solvents at a particular temperature. Thus, the distribution coefficient is defined as: $\frac{c_I}{c_{II}}=K_D$. According to Nernst distribution law, the concentration of a solute distributed between two phases at equilibrium and constant temperature bears a fixed ratio.

The formula of K_D is valid only when the molecular species are the same in two phases. If the solute molecule undergoes dissociation or association in any phases, the distribution law will not hold good. There are a few examples where the deviation is observed.

1) Distribution of a solute that dissociates in one of the phases: As an example, when oxalic acid (H_2Ox) is dissolved in ether and water, in the ether layer, oxalic acid molecules do not undergo dissociation, but it readily dissociates into ionized form in water. Assuming that the total concentration of oxalic acid in ether and aqueous layer be c_I and c_{II}, and the degree of dissociation of oxalic acid in water is α, we can write:

$$H_2Ox \leftrightharpoons H^+ + HOX^-$$

$$C_{II}(1-\alpha) \qquad \alpha C_{II}C_{II}$$

And the Nernst distribution law can be written as:

$$\frac{c_{II}(1-\alpha)}{c_I} = \text{constant} \tag{4.33}$$

2) Distribution of a solute when it associates in one of the phases: Such a case can be observed when benzoic acid (BH) is added to a mixture of benzene and water. Unlike above, here the benzoic acid assumes a monomeric form in the aqueous medium, but it readily dimerises in the benzoic acid medium, and equilibrium is established between dimers and monomers. This can be expressed using the following reaction:

$$2BH \rightleftharpoons (BH)_2$$

$$2c_I\alpha \qquad c_I(1-\alpha)$$

Like the above case, let the concentration of benzoic acid in benzene and water be c_I and c_{II}, respectively. α is the degree of association of the monomer to the dimer. If K is the equilibrium constant for the above, then K can be expressed as:

$$k = \frac{c_I(1-\alpha)}{4\alpha^2 c_I^2} = (4\alpha^2 c_I)^{-1} \quad \text{(Assuming } \alpha \ll 1\text{)}$$

$$\text{or, } \alpha = \frac{1}{2\sqrt{Kc_I}} \quad \textbf{(4.34)}$$

Thus, the concentration of monomer in benzene layer is:

$$2\alpha c_1 = \frac{2c_I}{2\sqrt{Kc_I}} = \text{Constant} = C_B \quad \textbf{(4.35)}$$

Thus, the distribution coefficient, K_D of benzoic acid between benzene and water is:

$$K_D = \frac{\text{The concentration of monomer in the aqueous layer}}{\text{The concentration of monomer in benzene layer}}$$

$$= \frac{c_W}{\sqrt{\frac{c_I}{c_W}}} = \frac{c_W}{\sqrt{C_I}} \quad \textbf{(4.36)}$$

For n-merisation, the above equation can be written in general form as:

$$\frac{c_{II}}{\sqrt[n]{c_I}} = \text{constant} \quad \textbf{(4.37)}$$

XII. TO STUDY THE PARTITION COEFFICIENT OR DISTRIBUTION COEFFICIENT OF IODINE BETWEEN CARBON TETRACHLORIDE AND WATER AND THE EQUILIBRIUM CONSTANT FOR THE REACTION BY PARTITION METHOD AND VERIFY THE NERNST DISTRIBUTION LAW

Chemicals and Apparatus

Distilled water, saturated solution of iodine, carbon tetrachloride (CCl_4), potassium dichromate ($K_2Cr_2O_7$), sodium thiosulphate ($Na_2S_2O_3$), potassium iodide (KI), starch, 4 stoppered bottle, mechanical shaker, 4 conical flasks, burette, 3 pipettes, measuring cylinder, volumetric flask (2 of 100 ml, 3 of 500 ml), separating funnel.

Theory

When iodine is dissolved in CCl_4 and water, according to different solubility in different solvents, iodine gets distributed into the two solvents. Since both iodine and CCl_4 are non-polar, so I_2 will be more soluble in CCl_4 than in water (soluble in very less quantity). At constant temperature, the partition coefficient according to Nernst's distribution law is:

$$K_D = \frac{c_{CCl_4}}{c_{H_2O}} \quad \textbf{(4.38)}$$

Here, C_{CCl4} and C_{H_2O} are the molar concentrations of iodine in CCl_4 and water, respectively. The reaction between KI and iodine follows the following equilibrium: $KI + I_2 \rightleftharpoons KI_3$. In an aqueous medium, applying the law of mass action, the equilibrium constant K_c of the reaction is given by:

$$K_C = \frac{c_{KI_3}}{c_{KI}c_{I_2}}.$$

K_c can be investigated by studying the distribution of iodine between CCl_4 and water, followed by a similar study for the distribution of iodine between the same organic solvent and an aqueous solution of KI.

Since distribution law is applicable to the species common to both layers, the concentration of free iodine in the aqueous KI layer can be determined from Nernst's equation (Equation 4.38). If the concentration of iodine in the CCl_4 layer is c_1 moles.lit^{-1} and the same in aqueous KI layer is c_2 moles.lit^{-1}, then, the concentration of free iodine in an aqueous layer is c_1/K_D moles/lit and concentration of KI_3 = total iodine concentration in aqueous layer-free iodine concentration $=(c_2-c_1/K_D) = c_2 - c_1/K_D$ moles.lit^{-1}

If the initial concentration of KI is c_3 moles.lit^{-1} and one mole KI is consumed to form the same amount of KI_3, thus, the concentration of free KI at equilibrium is:

$$\left[c_3 - \left(c_2 - \frac{c_1}{k_D}\right)\right] \text{ moles.lit}^{-1}$$

Thus, k_c can be expressed as:

$$k_c = \frac{\left(c_2 - \frac{c_1}{K_D}\right)}{\left[c_2 - \left(c_2 - \frac{c_1}{K_D}\right)\right]\left(\frac{c_1}{K_D}\right)} \text{ moles.lit}^{-1} \quad \textbf{(4.39)}$$

In the expression of the equilibrium constant, all the quantities are known, hence we can determine the equilibrium constant and the distribution coefficient using Equation (4.38) and Equation (4.39).

Procedure

1) Prepare 100 ml saturated solution of iodine in CCl_4, 0.1 (M) 100 ml $K_2Cr_2O_7$, 500 ml 0.1 $Na_2S_2O_3$ solution, 500 ml of 0.1 (M) KI solution. Prepare all the solutions in a volumetric flask. Prepare 1% of starch solution.

2) Prepare the following set of solution mixtures in 250 ml stoppered bottles:

Set	Vol. of Saturated I_2 in CCl_4 (ml)	Vol. of Pure CCl_4 (ml)	Vol. of Distilled Water (ml)
1	25	5	120
2	20	10	120

Shake both the bottles vigorously for an hour and then allow them to settle unless two distinct layers are observed (~30 min).

3) Standardise 0.1 (M) $Na_2S_2O_3$ against 0.1 (M) $K_2Cr_2O_7$ solution as follows:

Take 10 ml of $K_2Cr_2O_7$ using a pipette and pour into a conical flask.

Add 5 ml of conc. HCl, 100 ml of distilled water, and ~1 g of solid KI. Seal with stoppered and keep it in a dark place for 10 min.

Titrate it against 0.1 (M) $Na_2S_2O_3$ solution. When straw-yellow colour appears, add ~1 ml of starch solution. The solution will become very dark. Titrate till the appearance of light green colour. Note the burette reading (Table 4.18). Repeat the process until a precise value is obtained.

Calculate the strength of the solution and prepare 0.01 (M) 250 ml $Na_2S_2O_3$ solution from it.

4) Using a separating funnel, separate the layers. Ensure that the separation takes place properly. Pipette out 5 ml of organic layer (ensure that the pipette is dry) and pour it in a 100 ml conical flask. Add 10 ml of 0.1 (M) KI solution. Titrate it against 0.1 (M) thiosulfate solution. Note the titer value. Repeat titration until a good precise value is obtained.

Repeat the same for bottle no. 2.

5) Pipette out 25 ml of aqueous layer of bottle 1 and titrate it against 0.01 (M) thio (hypo) solution. Repeat the same for bottle no. 2.

6) Calculate the strength of I_2 of organic and aqueous layer for two bottles. Calculate K_D from Equation (4.38). In principle, K_D should be the same for both bottles. If this does not happen, take the average of the two different values of K_D (Table **4.19**).

Observation

Temperature = ... ° C

Table 4.17. Preparation of 100 ml 0.1 (M) $K_2Cr_2O_7$ solution.

Initial Mass of Weighing Bottle (g)	Final Mass of Weighing Bottle (g)	Mass of $K_2Cr_2O_7$ Transferred (g)	Strength of $K_2Cr_2O_7$ (M)
W_1	W_2	$W_2 - W_1$	-

Table 4.18. Standardisation of $Na_2S_2O_3$ solution against $K_2Cr_2O_7$ solution.

Obs. No.	Vol. of $K_2Cr_2O_7$ soln.	Vol. of hypo Solution reqd. (ml)	Precise Value	Strength of $Na_2S_2O_3$ (M)
1	10	-		-
2	10	-		-
3	10	-		-

Table 4.19. Titration result and determination of K_D.

Bottle No.	Vol. of Each Layer Taken (ml)		Vol. of 0.1 (M) $Na_2S_2O_3$ Solution for Organic Layer (V_{org}) (ml)	Vol. of 0.01 (M) $Na_2S_2O_3$ Solution for Aqueous Layer (V_{aq}) (ml)	C_{I_2} in Organic Layer (M/L)	C_{I_2} in Aqueous Layer (M/L)	KD	Average KD
	Organic	Aqueous						
1	5	25	....	....	....	....	....	....
2	5	25	....	....	....	....	....	....

Calculation

K_D for the system = ... at °C.

Instead of CCl_4, benzene, chloroform, hexane, or toluene can also be used.

(Tentative value for this particular case is 85 ± 5 at 25-30° C).

XIII. TO SHOW THAT BENZOIC ACID DIMERISES IN BENZENE AND DETERMINATION OF DIMERIZATION CONSTANT OF BENZOIC ACID IN BENZENE MEDIUM

Chemicals and Apparatus

Benzene, benzoic acid, oxalic acid, sodium hydroxide, phenolphthalein indicator, distilled water, stoppered bottles, conical flask, pipettes, burettes, measuring cylinder, mechanical shaker.

Theory

According to Nernst's distribution law at a constant temperature, the activities of a solute distributed between two phases at equilibrium bear a fixed ratio. The limitation of the law is that the law is valid only when the molecular species is the same in two phases. Thus, as a corollary, we can say that if a species starts aggregation or dissociation in one phase, the rule is no longer valid.

According to the Nernst distribution law, if a_1 and a_2 are activities of a solute in phase I and II, respectively at equilibrium at a given temperature, then, the distribution or partition coefficient of the solute between phase I and phase II, K_D is a ratio of a_I and a_{II}. Mathematically: $K_D = \frac{a_I}{a_{II}}$. For dilute solution, activity is replaced by concentration. Thus, K_D can be expressed as:

$$K_D = \frac{c_I}{c_{II}} \qquad (4.40)$$

When benzoic acid is shaken with two practically immiscible liquids, benzene, and water, then benzoic acid remains as monomer and almost un-dissociated in the aqueous phase, but dimerises easily in the benzene phase. If the molar concentration of benzoic acid in benzene and water be c_W and c_B, respectively, then, in that case, the above ratio $(\frac{c_W}{c_B})$ will no longer be valid. Instead, $\frac{c_W}{\sqrt{c_B}}$ = constant at a particular temperature. When the system achieves equilibrium, a couple of equilibria take place. Let us represent benzoic acid as BH.

(i) BA (aq.) $\rightleftharpoons$ BA (benzene);

$$K_D = \frac{(c_{BH})_B}{(c_{BH})_w} \quad \textbf{(4.41)}$$

(ii) 2 BH (benzene) $\rightleftharpoons$ $(BH)_2$ (benzene);

$$K = \frac{\left(c_{(BH)_2}\right)_B}{\left(c_{BH}^2\right)_B} \quad \textbf{(4.42)}$$

Here, K_D, and K are the distribution coefficients of benzoic acid between benzene and water; dimerisation constant of benzoic acid in benzene medium, respectively. $(c_{BH})_w$, $(c_{BH})_B$, $\left(c_{(BH)_2}\right)_B$ are the molar concentrations of benzoic acid in the water, benzene, and dimerized benzoic acid in benzene medium, respectively. The subscript w and B represent water and benzene, respectively.

If c_w^0 is the total molar concentration of BH in the aqueous phase and c_{wB}^0 is the total molar concentration of BH in benzene phase, then

$$c_w^0 = (c_{BH})_w \quad \textbf{(4.43)}$$

$$c_B^0 = (c_{BH})_B + 2\left(c_{(BH)_2}\right)_B \quad \textbf{(4.44)}$$

Thus, the ratio of c_B^0 and c_w^0 is obtained by dividing Equation. (4.44) by Equation. (4.43):

$$\frac{c_B^0}{c_w^0} = \frac{(c_{BH})_B}{(c_{BH})_w} + \frac{2\left(c_{(BH)_2}\right)_B}{(c_{BH})_w}$$

$$= \frac{(c_{BH})_B}{(c_{BH})_w} + \frac{2\left(c_{(BH)_2}\right)_B}{(c_{BH})_w} \times \frac{\left(c_{BH}^2\right)_w}{\left(c_{BH}^2\right)_w} \times (c_{BH})_w \quad \textbf{(4.45)}$$

$$\frac{c_B^0}{c_W^0} = K_D + 2KK_D^2 c_W$$

From Equation. (4.41) and (4.42)

Thus, a plot of $\frac{c_B^0}{c_W^0}$ *Vs*. c_W is a straight line with a slope of 2 K K_D^2 and intercept K_D (See plot below). The quantities K and K_D can be calculated experimentally also.

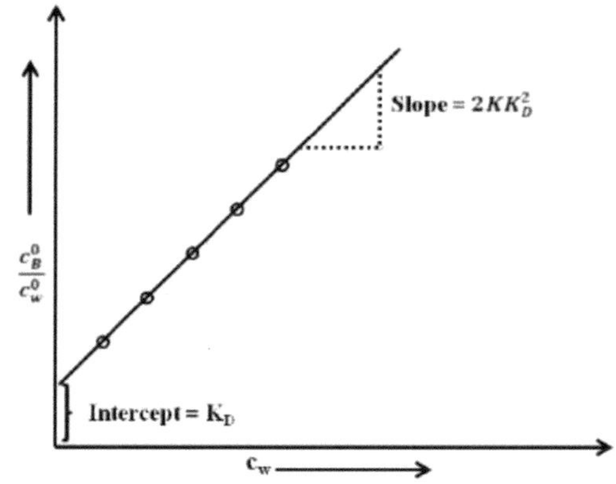

Fig. (4.6). Plot of ratio of the total molar concentration of benzoic acid in benzene and water *Vs*. the concentration of benzoic acid in water.

In this study, we focus on two things:

1) To show that the ratio of the concentration of benzoic acid in water and benzene is not constant; rather the ratio of the concentration of benzoic acid in water and the square root of concentration of benzoic acid in benzene is constant.

2) To evaluate the equilibrium constant of dimerisation of benzoic acid in benzene and the distribution coefficient of benzoic acid in benzene.

Procedure

1) Prepare a 200 ml saturated solution of benzoic acid in benzene.

2) Prepare different sets of solutions of benzoic acid in benzene by following Table 4.20. Shake all the sets of bottles for an hour and then allow them to settle for half an hour.

3) Prepare ~0.1 (M) 500 ml NaOH solution (Dissolve ~2 g of NaOH pellets in 500 ml water) and standardise the same against 0.05 (M) oxalic acid solution (exact strength should be known as in Table **4.21**. Get the exact strength of the NaOH solution (Table **4.22**). Dilute it ten times to prepare the exact 0.01 (M) NaOH solution (Make in a volumetric flask).

4) Take anyone from the above sets of bottles (say set 1) and separate the two layers using a separating funnel and collect the two liquids in different conical flasks and put a cover on them. Pipette out 25 ml of the aqueous layer in a 125 ml conical flask and titrate it against 0.01 (M) NaOH solution using phenolphthalein indicator. Repeat titration unless getting a good agreement of titer value (Table **4.23**).

To titrate the benzene layer, take 5 ml of benzene layer and pour into 25 ml distilled water in a 125 ml conical flask. Titrate it following the above procedure (Table **4.24**).

5) Repeat step 4 for all the other bottles.

Observation and Tables

Temperature =.... ° C

Table 4.20. Preparation of solutions.

Set	Vol. of Benzoic Acid in Benzene (ml)	Vol. of Pure Benzene (ml)	Vol. of Distilled Water (ml)
1	50	0	100
2	40	10	100
3	30	20	100
4	20	30	100
5	10	40	100

Table 4.21. Preparation of 0.05 (M) oxalic acid solution.

Mass of Empty Weighing Bottle (g)	Mass of Weighing Bottel (g)	Mass of Oxalic Acid Transferred (g)	Strength of Oxalic Acid Solution (0.05(M))
W_1	W_2	(W_2-W_1)	$(W_2-W_1)/$ (126.06544)

Table 4.22. Standardisation of NaOH solution against the standard oxalic acid solution.

Strength of Oxalic Acid (M)	Volume of Oxalic Acid solution taken (ml)	Volume of NaOH solution required (ml)	Most Precise Value of NaOH volume (ml)	Strength of NaOH solution (M)
-	10 ml	....	-	-
		....		
		....		

Table 4.23. Titration of the aqueous layer.

Set no.	Vol. of aq. Layer (ml)	Vol. of 0.01 (M) NaOH Soln. Required (ml)	Most precise Value of NaOH Volume (ml)	c_W (Moles/lit)
1	25	....	-	-
		....		
		....		
2	25	....	-	-
		....		
		....		
3	25	....	-	-
		....		
		....		
4	25	....	-	-
		....		
		....		
5	25	....	-	-
		....		
		....		

Table 4.24. Titration of Benzene Layer.

Set no.	Vol. of Benzene Layer (ml)	Vol. of 0.1 (M) NaOH soln. Required (ml)	Most Precise Value of NaOH Volume (ml)	c_B (Moles/lit)
1	25	....	-	-
		....		
		....		

2	25	....	-	-
		....		
		....		
3	25	....	-	-
		....		
		....		
4	25	....	-	-
		....		
		....		
5	25	....	-	-
		....		
		....		

Table 4.25. Table for calculation of $\frac{c_W}{c_B}$ and $\frac{c_W}{\sqrt{c_B}}$.

Set no.	$\frac{c_W}{c_B}$	$\frac{c_W}{\sqrt{c_B}}$ (mole$^{1/2}$lit$^{-1/2}$)
1	-	-
2	-	-
3	-	-
4	-	-
5	-	-

Plot $\frac{c_W}{c_B}$ *vs.* $\mathbf{c_W}$ (Fig. **4.6**) From graph intercept = K_D and slope= $K\,K_D^2$. Since all the quantities are known here, K and K_D can be calculated easily.

FURTHER READING

[1] Laidler, K. J. "Chemical Kinetics" 3rd Edition, Pearson Education, Ottawa, Canada, 1997.
[2] Kapoor, K. L. "A text Book of Physical Chemistry" McMillan Publishers. 2000. Vol. 2.
[3] Rakshit, P. C. "Physical Chemistry" 7th Edition, Sarat Book House, Calcutta, India, 2014.
[4] Levine, I. N. "Physical Chemistry" Tata Mc. Graw Hills, New Delhi, India, 2011.
[5] Atkins, P. W., de Paula, J. "Atkins' Physical Chemistry" 11th Edition, Oxford University Press, 2018.
[6] Glasstone, S. "Textbook of Physical Chemistry" 2nd edition, Macmillan, 1958, Digitized 2009.
[7] Palit S. R.; De, S. K. "Practical Physical Chemistry" Science Book Agency, Kolkata, 1974.
[8] Daniels, F.; Mathews, J. H.; Williams, J.W.; Bender, P.; Alberty, R. A. "Experimental Physical Chemistry" McGraw-Hill Book Company INC, New York, Toronto-London, 1956.
[9] Athawale, V. D.; Mathur, P. "Experimental Physical Chemistry" New Age International (P) Limited Publishers, New Delhi, 2011.

CHAPTER 5

Phase Diagram

GENERAL DISCUSSION

The phase rule is used for the quantitative treatment of systems in equilibrium. It helps to predict the conditions that must be specified for a system to exhibit equilibrium. One should know terms like phase, components, and degree of freedom before going through the phase rule.

A phase (P) is defined as any homogeneous and physically distinct part of a system that is bounded by a surface and is mechanically separable from other parts of the system.

The number of components (C) of a system at equilibrium is defined as the smallest number of independently variable constituents by means of which the composition of each phase can be expressed either directly or in terms of the chemical equation.

Let us take an example of ice, water, and vapour in equilibrium:

$$\text{ice} \leftrightharpoons \text{liquid water} \leftrightharpoons \text{vapour}$$

Here, at equilibrium three phases can coexist, but the number of components is 1 (As only water is considered). In 1876 A.D., J. W. Gibbs proposed a rule to make a correlation between degrees of freedom, component, and phase, which is known as the phase rule. The phase rule is defined as F = C-P+2. Here, F is degrees of freedom. The number 2 denotes 2 independent variables, *viz.* temperature, pressure. The degree of freedom of a system is defined as the number of independent variables. Thus, in the above example using the phase rule we can say that the number of degrees of freedom is zero, (C = 1, P = 3). Thus, it is an invariant system. So, based on the value of F, we can say invariant (F = 0), univariant (F = 1), bivariant (F = 2), trivariant (F = 3) and so on.

Although, the phase diagram is applicable even for solid solutions (alloys). Here we shall demonstrate only the liquid mixtures.

Charu Arora and Sumantra Bhattacharya

I. TO STUDY THE CRITICAL SOLUTION TEMPERATURE OF A BINARY MIXTURE

Chemicals and Apparatus

Phenol, distilled water, thermostat/water bath, test tube, thermometer.

Theory

Phenol and water are not miscible completely; rather they are partially miscible at room temperature. When a small quantity of phenol is added to water, the substance will dissolve completely. If the addition of phenol is continued for a certain time it will be observed that no further dissolution takes place and the two liquid layers separate. Each layer is a saturated solution of one in the other. These two layers in equilibrium are known as conjugate solutions. The upper layer is a saturated solution of phenol and water and the lower one is water in phenol.

Mutual solubility of phenol and water increases with an increase in temperature which causes the concentration of phenol as well as that of water in phenol. At a particular temperature, two conjugate solutions convert into a homogeneous solution. The particular temperature is known as the critical solution temperature (CST). This temperature is the characteristic of a particular system and impurities of the system influenced the shift of CST. If pure water and phenol are taken, at 66.4° C the CST is observed. This is known as upper critical solution temperature (UCST). The plot of the composition of phenol added to water *vs.* temperature is plotted as in Fig. (**5.1**).

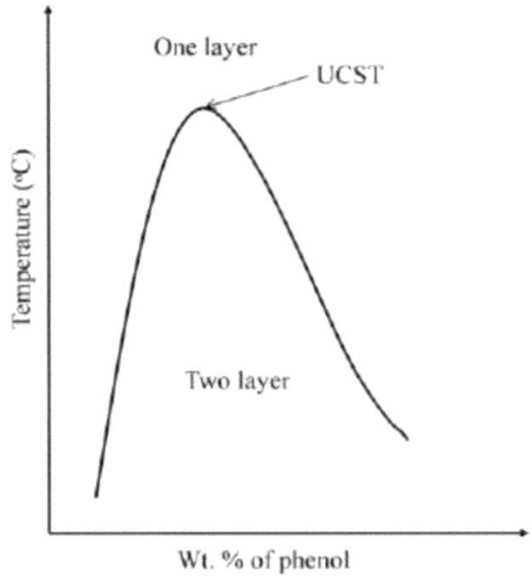

Fig. (5.1). UCST of phenol water system.

At any temperature above UCST, phenol-water is miscible in all proportions. Outside the curve, the system is completely homogeneous, *i.e.*, outside the curve, only one layer persists. Below the curve always two layers will separate and the curve gives the composition of the two conjugate solutions consisting of two layers. Throughout the experiment, pressure is kept fixed. Thus, the phase rule has been reduced to F=C-P+1.

For a homogeneous phase of two-component liquids, $F = 2 - 2 + 1 = 2$, *i.e.*, degrees of freedom is 2, implying that the system is bivariant. Thus, for homogeneous solutions both composition and temperature have to be defined.

For the heterogeneous equilibrium with two conjugate solutions, $F = 2 - 2 + 1 = 1$. This implies that the system would behave as a univariant. So, either temperature or composition is fixed and another parameter will be known automatically from the diagram. At CST composition of the two conjugate solutions be the same, *i.e.*, $F = 2 - 2 + 1 - 1 = 0$. Thus, the point is fixed. For a particular pair of partially miscible liquids, the value is fixed.

Procedure

1) Note down the room temperature and get the density of water from the table of the variation of density of water with respect to temperature.

2) Weigh ~ 2 g of phenol accurately (Table **5.1**, density of phenol = 1.07 g/cc at 298 K) and transfer it to the hard glass test tube.

3) Fit the hard glass test tube (tt) with a holder fixed to a burette stand. Dip a thermometer inside the tt. Immerse the bottom of the tt into distilled water filled in a 250 ml beaker which is kept on an asbestos board over the tripod stand.

4) Add 1 ml of distilled water into the hard glass tt by using a pipette to the phenol. Stir the phenol-water solution well with the stirrer so that the mixture becomes turbid. Heat the bath slowly until the mixture becomes clear suddenly at a particular temperature. Note the temperature of the disappearance of turbidity. Remove the flame and allow the liquid to cool down slowly. While cooling down, turbidity reappears. Note the temperature. Take the average of two temperatures (Table **5.2**).

5) Add 1 ml of water to the system and repeat step 3 and keep on adding the same volume of water to the system and repeat. Take at least 10-15 readings.

6) Calculate the weight percentage of phenol for each step. Plot mean temperature *Vs* . wt.% of phenol using the data in Table **5.3**. Maxima of the graph correspond to UCST. Draw a perpendicular line on both the axis. The point corresponds to % of phenol and temperature at UCST.

7) Try the experiment by starting with 2 ml of water and add 1 ml of phenol at a time. Follow the above procedure. Compare UCST, temperature, and composition at UCST of both.

Observation and Results

Temperature =.... ° C

Table 5.1. Weighing of phenol.

Initial Mass of Phenol+wt. of Weighing Bottle (g)	Mass of Empty Weighing Bottle (g)	Mass of Phenol Transferred (g)
W_1	W_2	$(W_1 - W_2)$

Table 5.2. Recording of miscibility temperature.

Observations	Vol. of Water Added (ml)	Miscibility Temperature (° C)		
		Turbidity Disappearance	Turbidity Reappearance	Mean
1	1	....	....	....
2	2 (1+1)	....	....	....
3	3 (2+1)			....
4	4 (3+1)			....
5	5 (4+1)			....
...	....	....	....	....
15	....	....	....	....

Table 5.3. Table for graph plotting.

Weight Percentage of Phenol (%)	Mean Miscibility Temperature (° C)

Conclusion

From the plot of weight % of phenol *Vs* . miscibility temperature

Critical solution temperature (CST) =…° C,

Critical composition = …. % of phenol by weight

(II) TO STUDY THE CRITICAL SOLUTION TEMPERATURE OF A TERNARY MIXTURE

Chemicals and Apparatus

Benzene (Bz), distilled water, glacial acetic acid (AcOH), graduated pipette, burettes, 125 ml conical flask (3 pieces), 1 no. of 250 ml conical flask.

Theory

The phase rule is: F = C – P + 2. Here, F, C, and P are degrees of freedom, component, and phase of a system, respectively. 2 appears for two parameters, like, temperature (T) and pressure (P). Thus, for an isothermal and isobaric three-component system the phase rule reduces to F = 3 - P. The composition of such a system can be expressed in terms of the coordinates of an equilateral triangle, where each side corresponds to a pure component. Thus, any point, say X inside the graph can give information about the composition of three components (Fig. **5.2**). Benzene and water are essentially immiscible (better to say as sparingly miscible); hence two layers will be observed on mixing. Acetic acid is soluble in both benzene and water. Thus, on its addition to a mixture of benzene and water, acetic acid distributes itself between these layers. The composition of the layers changes as more acetic acid is added. Under these conditions, P = 2, and the system is completely defined by the composition. Thus, if the points corresponding to these compositions are plotted, a smooth curve can be drawn indicating the limits of miscibility of the water-benzene mixture in acetic acid.

It is notable that as more and more acetic acid is added, the water and benzene solution becomes more and more miscible until complete miscibility is achieved.

This solubility curve is known as the bimodal curve. Any mixture within the area enclosed by this bimodal curve and the base of the triangle will resolve itself into two liquid layers and any mixture outside the area will form only one liquid layer. The position of the curve changes with temperature.

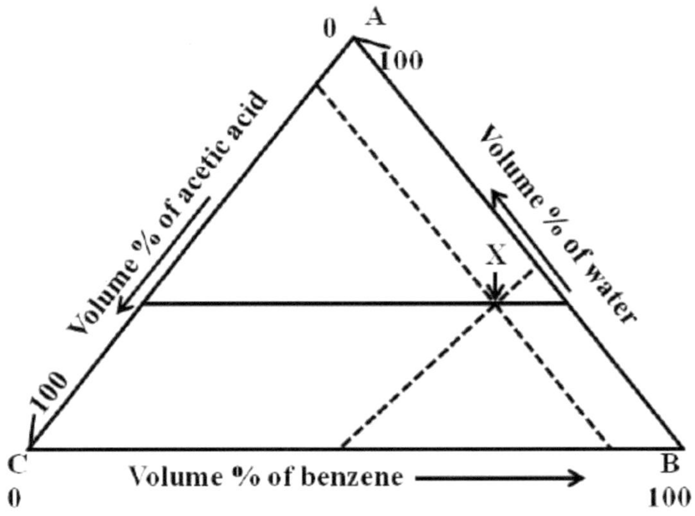

Fig. (5.2). Phase diagram plot of a three-component system.

Procedure

1) Take 5 ml of benzene in a 125 ml clean and dry conical flask. 1 ml of distilled water is added using a pipette. Glacial acetic acid is slowly added from another burette unless a clear homogeneous solution has just been obtained (shake vigorously while mixing). Record the volume of acetic acid added.

2) Add 5 ml of benzene to the mixture and the mixture adds 1 ml of acetic acid as mentioned in step 1.

3) Repeat step 2 at least 10 times and record the data using Table **5.1**. Name the set of the experiment as SET I (Table **5.4**).

4) In another set of experiments, say SET II, take 5 ml of benzene in a 125 ml clean and dry conical flask and add successively 3, 4, 5, 6, 7... ml portions of water and titrate with glacial acetic acid solution after every addition. Take at least 10 readings. Record the data in Table **5.5**.

5) In the last series of the experiment (SET III), take 2 ml of benzene in a 250 ml clean and dry conical flask and is treated successfully with eight 25 ml portions of water and titrate with acetic acid after each addition (Table **5.6**).

6) Calculate volume percentage (V/V) of benzene, water, and acetic acid for each homogeneous solution and plot on a triangular graph paper, as shown above in Fig. **(5.2)** and record in Table **5.7**.

7) Plot another graph of volume of benzene per 100 ml of acetic acid against a volume of water per 100 ml of acetic acid using Table **5.8** and indicate the homogeneous and heterogeneous phases on the graph paper (Fig. **5.3**).

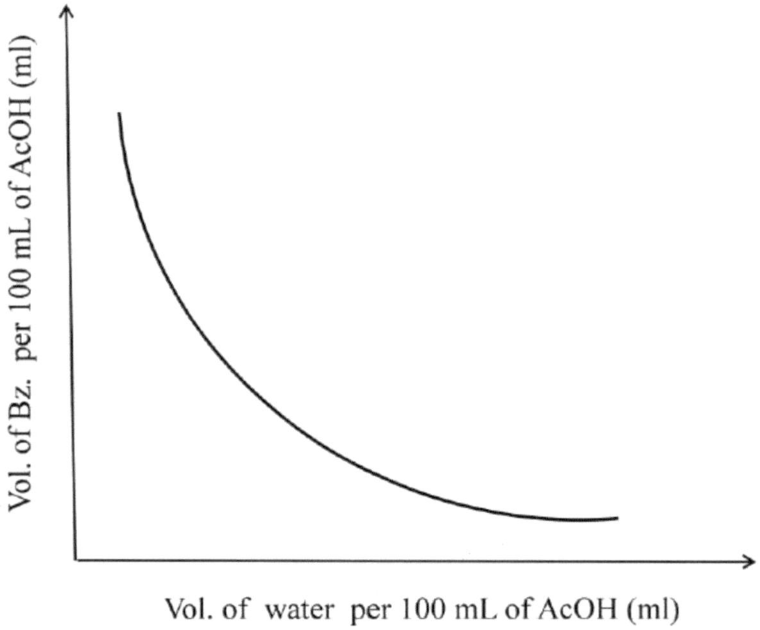

Fig. (5.3). 2-dimensional plot of the volume of benzene *vs.* volume of water per 100 ml of acetic acid.

Observation

Temperature = …. ° C

Table 5.4. Composition of SET I.

Volume of Water (ml)	Volume of Benzene (ml)	Minimum Volume of Acetic Acid Required (ml)
1	5	-----
1	10	-----
1	15	-----
1	20	-----
1	25	-----
1	30	-----
1	35	-----
1	40	-----
1	45	-----
1	50	-----

Table 5.5. Composition of SET II.

Volume of Water (ml)	Volume of Benzene (ml)	Minimum Volume of Acetic Acid Required (ml)
3	5	-----
6	5	-----
9	5	-----
12	5	-----
15	5	-----

18	5	-----
21	5	-----
24	5	-----
27	5	-----
30	5	-----

Table 5.6. Composition of SET III.

Volume of Water (ml)	Volume of Benzene (ml)	Minimum Volume of Acetic Acid Required (ml)
25	2	-----
50	2	-----
75	2	-----
100	2	-----
125	2	-----
150	2	-----
175	2	-----
200	2	-----

Table 5.7. Determination of percentage of each component for each homogeneous solution.

SERIES I			SERIES II			SERIES III		
Vol % of Bz	Vol % of H_2O	Vol % of AcOH	Vol % of Bz	Vol % of H_2O	Vol % of AcOH	Vol % of Bz	Vol % of H_2O	Vol % of AcOH

Table 5.8. Table for a two-dimensional graph.

SERIES I		SERIES II		SERIES III	
Vol. of Bz per 100 ml of AcOH (ml)	Vol. of H_2O per 100 ml of AcOH (ml)	Vol. of Bz per 100 ml of AcOH (ml)	Vol. of H_2O per 100 ml of AcOH (ml)	Vol. of Bz per 100 ml of AcOH (ml)	Vol. of H_2O per 100 ml of AcOH (ml)

Calculation

Let us consider SERIES I

Vol. of water = 1 ml

Vol. of Bz. = 5 ml

Vol. of AcOH required = x ml

Thus, Vol. % of water = (1*100)/(6+x)

Vol. % of Bz = (5*100)/(6+x)

Vol. % of AcOH = (x*100)/(6+x)

Similarly,

Vol. of Bz per 100 ml of AcOH = 5*100/x ml

Vol. of water per 100 ml of AcOH = 1*100/x ml

Two-dimensional plot is plotted

FURTHER READING

[1] Rakshit, P. C. "Physical Chemistry" 7th edition, Sarat Book House, Calcutta, India, 2014.
[2] Palit S. R.; De, S. K. "Practical Physical Chemistry" Science Book Agency, Kolkata, 1974.
[3] Daniels, F.; Mathews, J. H.; Williams, J.W.; Bender, P.; Alberty, R. A. "Experimental Physical Chemistry" McGraw-Hill Book Company INC, New York, Toronto-London, 1956.
[4] Athawale, V. D.; Mathur, P. "Experimental Physical Chemistry" New Age International (P) Limited Publishers, New Delhi, 2011.

CHAPTER 6

Adsorption

INTRODUCTION

The term Adsorption was first coined in 1881 by a German physicist named Heinrich Kayser. Adsorption is purely a surface phenomenon, where particles are attached to the top layer of material. It normally involves the molecules, atoms, even ions of a gas, liquid, or solid in a dissolved state that is attached to the surface.

Adsorption is mainly a consequence of surface energy. Generally, the surface particles which can be exposed partially tend to attract other particles to their site. Interestingly, adsorption phenomena is widely found in various natural processes and finds its use in many industrial applications. Some examples are summarized below:

1) Air Pollution Masks

These consist of silica gel or activated charcoal powder, when dust or smoke are passed through them, those particles get adsorbed on the surface of these materials. Additionally, these types of masks are effective to protect microbes also.

2) Separation of Noble Gases by Dewar's Flask Process

A mixture of noble gases of second to fourth row (Ne, Ar, Kr) are passed through Dewar's flask in presence of heated coconut charcoal. Argon and Krypton gas get adsorbed leaving Neon.

3) Purification of Water

By the addition of alum stone to the water, impurities get adsorbed on the alum, and water gets purified.

4) Removal of Moisture and Humidity

Moisture in the air can be removed by placing silica gel on which water molecules get adsorbed.

Charu Arora and Sumantra Bhattacharya

5) Adsorption Chromatography

It is used to separate pigments and hormones.

6) Ion Exchange Method

In this method of removing the hardness of water, calcium and magnesium ions get adsorbed on the surface of the ion exchange resin.

7) In Metallurgy

In the froth floatation process of concentration of ore, the particle gets adsorbed on the froth.

Difference between adsorption and absorption: Absorption is a process in which a fluid is dissolved by liquid or a solid (absorbent). Adsorption is a process in which atoms, ions or molecules from a substance adhere to the surface of adsorbent. So be careful about using these two words. Do not get confused!

I. STUDY THE ADSORPTION OF ACETIC ACID ON CHARCOAL AND PROVE THE VALIDITY OF FREUNDLICH/LANGMUIR ADSORPTION ISOTHERM

Chemicals and Materials Required

0.5 (N) acetic acid (AcOH), 0.1 (N) sodium hydroxide (NaOH), charcoal, phenolphthalein indicator, oxalic acid (Ox), burette, pipette, reagent bottles, conical flask, and shaker.

Theory

Adsorption is a surface phenomenon. Adsorption can take place between any states of matter (gas, liquid, solid) on solid. The substance which gets adsorbed is called adsorbate and the surface of the solid on which it is adsorbed is known as adsorbent. Depending upon the nature of surface forces involved adsorption can be classified as physisorption and chemisorption. Physisorption or physical adsorption mainly arises because of very weak van der Waals' interaction between adsorbent and adsorbate. On the other hand, the interaction between them is much stronger in chemisorption. The heat of adsorption lies in the range of 80-420 kJ/mol with high activation energy for chemisorption. Usually, it is irreversible. While for physical adsorption, the heat of adsorption is less than 20 kJ/mol, and activation energy is often less than 5 kJ. Physisorption is multilayer, while chemisorption is a monolayer.

The amount adsorbed (a) is dependent on pressure (P) and temperature (T). Hence, a can be expressed as a function of pressure and temperature, a=f (P, T). A plot of P and a, keeping the temperature constant is known as adsorption isotherm.

> In an **isothermal** process, temperature is constant. If pressure is fixed is called **isobaric** and if volume is uniform, then it is known as **isochoric**.

Freundlich Adsorption Isotherm

Freundlich (1909 AD) proposed an empirical relationship: $\frac{x}{m} = kc^{\frac{1}{n}}$. Here, x is the amount of solute adsorbed and m is the amount of adsorbed material. c is the equilibrium concentration of adsorbate in the solution, k is a constant, depends on the nature of adsorbent and adsorbate. n is another constant, dependent on the nature of adsorbate. The value of 1/n lies between 0 to 1. On taking the logarithm of the above equation, we have:

$$\log\left(\frac{x}{m}\right) = \log k + \frac{1}{n}\log c \qquad \textbf{(6.1)}$$

If we plot the left-hand side of the equation along the Y-axis and log c along the X-axis, we get a straight-line having slope 1/n, and log k is the intercept along the x-axis (Fig. **6.1**).

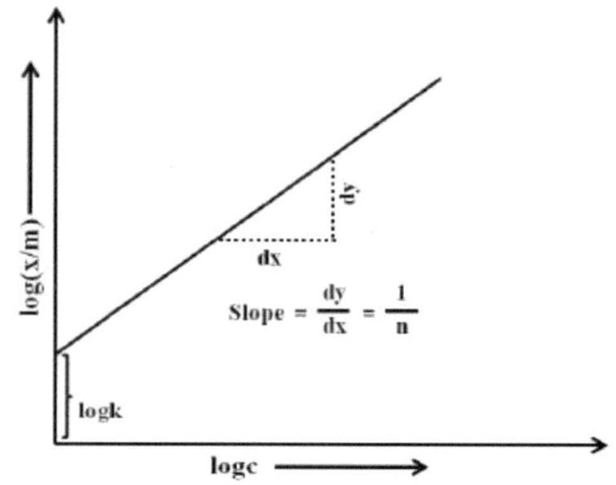

Fig. (6.1). Plot logarithm of concentration *vs*. logarithm of (x/m) (Freundlich adsorption isotherm).

Langmuir Adsorption Isotherm

Langmuir proposed an equation that establishes the relation between the amount adsorbed and the concentration for a unimolecular layer, known as Langmuir adsorption isotherm.

Suppose a material (A) is adsorbed on another solid material (S). The adsorbed complex is A_{ad}. Then, we can express:

$$A + S \overset{K_{eq}}{\rightleftarrows} A_{ad}$$

If c is the concentration of adsorbent. Suppose x is the amount of adsorbate is adsorbed by m gram of adsorbent, then according to the Langmuir adsorption isotherm we can write:

$$\frac{x}{m} = \frac{k_{eq}c}{1 + k_{eq}c} \quad \textbf{(6.2)}$$

This expression can also be written as:

$$\frac{c}{x/m} = \frac{1}{k_1 k_2} + \frac{c}{k_2} \quad \textbf{(6.3)}$$

If values of $\frac{c}{x/m}$ are plotted as ordinate against c as abscissa, we get a straight line, having slope $1/k_2$ and intercept = $1/(k_1k_2)$ on the ordinate (Fig. **6.2**).

A charcoal powder has a very fine structure and hence has a very high surface area. For this reason, charcoal can adsorb a significant amount of several gases and liquids. In addition to that getting charcoal is very cheap and nontoxic. So, charcoal can be used as an ideal material to study the phenomena of adsorption in the laboratory.

Procedure

1) Prepare 0.5 (N) Oxalic acid, 0.5 (N) AcOH, and 0.1 (N) NaOH solutions and standardize them using standard oxalic acid. Take six stoppered reagent bottles and prepare the following solutions in each bottle (Table **6.1**). For every bottle take 1.0 g (m) of charcoal:

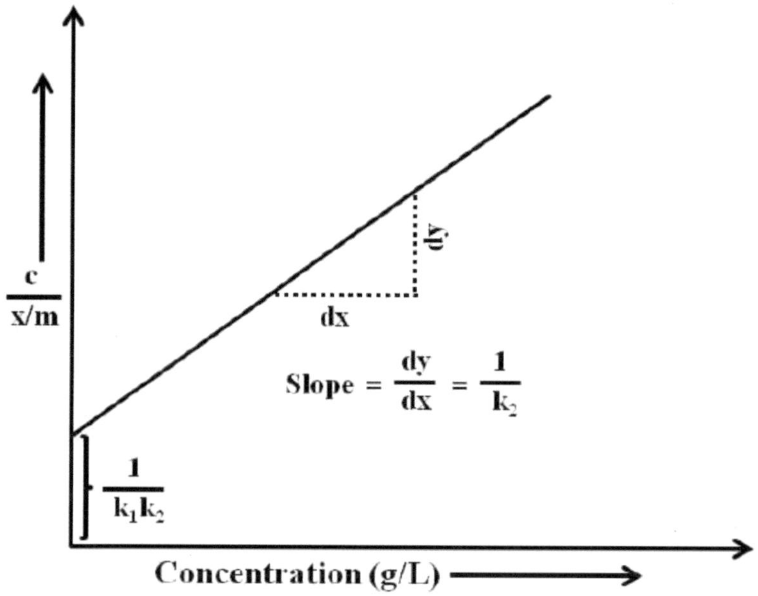

Fig. (6.2). Plot of Langmuir adsorption isotherm.

Table 6.1. Preparation of acetic acid solution.

Bottle	0.5 (N) Acetic Acid (ml)	Distilled Water (ml)
1	50	0
2	40	10
3	30	20
4	20	30
5	10	40
6	5	45

2) Shake all the bottles thoroughly well for at least 1 hour on a mechanical shaker. Allow them to settle. Filter each solution using filter paper and collect the filtrate in already numbered beakers. While filtering, discard the first 5 ml of each filtrate. Because some acid gets absorbed by the filter paper in the initial fraction of filtrate, leads to experimental error (error of method). Pipette out 10 ml of each filtered solution in a conical flask and titrate it against 0.1 (N) NaOH solution, using phenolphthalein indicator. Repeat the titration twice for each solution and consider the most precise value (Never take an average reading). Titrate the stock solution of AcOH (10 ml) with 0.1 (N) NaOH (Table **6.2**).

Observations

Temperature =……°C

10 ml of stock solution AcOH ≡ x ml 0.1 (N) NaOH

Table 6.2. Analysis of acid adsorbed.

Bottle	Initial Concentration of Acid Before Adsorption (c_0) (Suitable Unit)	Equilibrium Concentration of Acid after Adsorption (c_e) (Suitable Unit)	Amount of Acid Adsorbed (c_0- c_e) ≡ ml of NaOH
1	x	……	……
2	4x/5	……	……
3	3x/5	……	……
4	2x/5	……	……
5	x/5	……	……
6	x/10	……	……

Calculations

x = (c_0- c_e). Calculate x/m for each bottle and find the value of log(x/m), as well as log c_e for each bottle. Plot a graph of log (x/m) and log c_e as abscissa. A straight line with slope 1/n is observed proving the validity of Freundlich adsorption isotherm.

To prove the validity of the Langmuir adsorption isotherm, the values of $\frac{c_e}{x/m}$ are plotted as ordinate against c_e as abscissa. A straight line will be observed if Langmuir adsorption isotherm is obeyed.

Result

The validity of Freundlich and Langmuir isotherm has been tested and verified for adsorption of AcOH on charcoal.

II. TO STUDY THE ADSORPTION OF IODINE FROM ALCOHOLIC SOLUTION ON CHARCOAL

Chemicals and Materials

Iodine, ethanol, sodium thiosulphate ($Na_2S_2O_3$ or hypo solution), burette, pipette, volumetric flask (250 ml) reagent bottles, conical flask, thermostat, and shaker.

Theory

Same as **I.**

Procedure

1) Dissolve 32 g of iodine in ethanol in a 250 ml volumetric flask and make the volume up to the mark. The strength of iodine solution is ~0.5 (M)

2) Take 6 differently labeled stoppered reagent bottles and pour accurately weighed 1.0 g activated charcoal powder in each of them.

3) Add iodine solution and pure ethanol in each of the six bottles in the following manner (Table **6.3**).

Table 6.3. Preparation of iodine solution.

Bottle	0.5 (N) Iodine Solution (ml)	Pure Ethanol (ml)
1	50	0
2	40	10
3	30	20
4	20	30

5	15	35
6	10	40

4) Shake all the bottles thoroughly well for at least 1 hour on a mechanical shaker and keep them in a thermostat.

5) Prepare 500 ml 0.2 (M) hypo solution and standardize it. Titrate the hypo solution against 10 ml iodine to determine the exact concentration of the stock solution of iodine. The starch indicator is added as mentioned in chapter 3.

6) Filter the contents of each bottle using filter papers and collect the filtrate in a cleaned labeled conical flask. Reject 1st 5 to 10 ml portion of filtrate from each bottle to minimize the error of the method.

7) Titrate 5 ml each from bottle 1 and 2, 10 ml from bottle 3 and 4, and 20 ml from two remaining bottles against standard hypo solution as is done with the stock solution.

8) Calculate equilibrium concentration in each bottle.

9) Amount of iodine adsorbed by charcoal is calculated similarly as was done in the previous experiment.

10) Tabulate the observations and results and test the validity of Freundlich and Langmuir adsorption isotherms as was done in the previous experiment.

FURTHER READING

[1] Laidler, K. J. "*Chemical Kinetics*" 3rd Edition, Pearson Education, Ottawa, Canada, **1997**
[2] Kapoor, K. L. "*A Text Book of Physical Chemistry*" Mc. Graw Hill, New Delhi, India, **2019**, Vol 7.
[3] Rakshit, P. C. "*Physical Chemistry*" 7th Edition, Sarat Book House, Calcutta, India, **2014**.
[4] Levine, I. N. "*Physical Chemistry*" 6th Edition, Tata Mc. Graw Hills, New Delhi, India, **2011**
[5] Atkins, P. W.; de Paula, J. "*Atkins' Physical Chemistry*" 11th Edition, Oxford University Press, **2018**.
[6] Glasstone, S. "*Textbook of Physical Chemistry*", 2nd Edition, Macmillan, **1958,** Digitized **2009.**
[7] Palit S. R.; De, S. K. "*Practical Physical Chemistry*" Science Book Agency, Kolkata, **1974**.
[8] Daniels, von F.; Mathews, J. H.; Williams, J.W.; Bender, P.; Alberty, R. A.; "*Experimental Physical Chemistry*" McGraw-Hill Book Company INC., New York-Toronto-London **1956**.
[9] Athawale, V. D.; Mathur, P. "*Experimental Physical Chemistry*" New Age International (P) Limited Publishers, New Delhi, **2011**.

CHAPTER 7

Instruments Based on Optical Measurements

REFRACTOMETRY

Introduction

When a beam of light passes from one medium to another medium it deviates from its path. This phenomenon is known as refraction. If it passes from a less rare medium to a denser medium the beam refracted towards the normal (Fig. **7.1**).

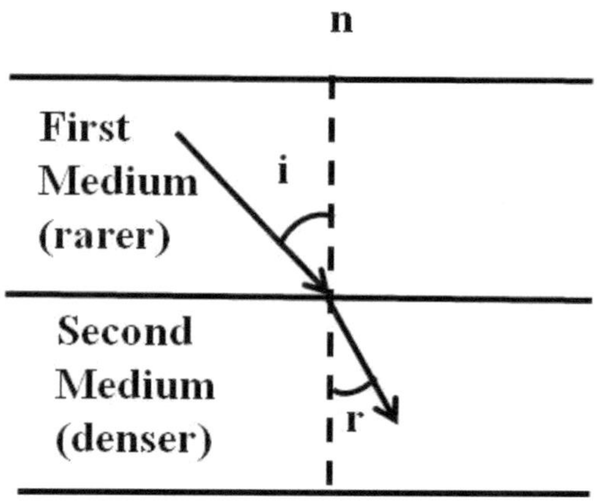

Fig. (7.1). Phenomena of refraction.

In this case the angle of incidence (angle between normal and incident light), i is greater than the angle of refraction (angle between normal and refracted beam), r. The refractive index of the second medium (n) with respect to first can be given by Snell's law of refraction.

$$n = \frac{\sin i}{\sin r} \tag{7.1}$$

The value of the refractive index can also be given by

$$n = \frac{\text{Velocity of light in vacuum}}{\text{velocity of light in the medium}} \tag{7.2}$$

Charu Arora and Sumantra Bhattacharya

The Refractive index of a liquid depends on temperature as well as the wavelength of light used.

Specific Refractivity

Specific refraction or specific refractivity is expressed as:

$$R = \frac{1}{d} \times \frac{n^2-1}{n^2+2} \quad \textbf{(7.3)}$$

Here R is specific refractivity and it is independent of temperature, d is density.

Molar Refractivity

Molar refractivity can be obtained by multiplying the specific refractivity by the molecular weight of the medium. It can be expressed as:

$$R_M = R \times M = \frac{M}{d} \times \frac{n^2-1}{n^2+2} \quad \textbf{(7.4)}$$

Molar refractivity is the constitutive and additive property and we have refractivity due to atoms (atomic refractivity) as well as refractivity due to structural factors (structural refractivity).

Description of Abbe Refractometer

Many refractometers are available for rapid and accurate determination of the refractive index. For small quantities of liquids, we can easily use the Abbe refractometer (Fig. **7.2**) in the laboratory. It consists of two flint glass prisms, illuminating prism (lower prism), and a measuring prism (Upper prism). The surface of the measuring prism is finely polished while that of the illuminating prism is finely ground. The two prisms are fixed in a metal casing. The prisms are jacketed so that they can be maintained at constant temperature by circulating water. The prisms are rotated using a movable arm that carries reading glass. A thin layer of liquid can be placed between two prisms. The position of the borderline of a total reflection is observed through a telescope and by turning the movable arm, it can be made to coincide with the intersection of the cross wire in the telescope. The arc is graduated in such a way that it gives directly the value of the refractive index with an accuracy of 0.001. Ordinary light can also be used while working with the Abbe refractometer. The telescope is provided with the dispersion compensator, for this purpose. The dispersion compensator consists of an Amici prism mounted one over the other and can be rotated in opposite direction by turning the milled head. When we put a drop of liquid on the surface of the illuminating prism and clamp it with a measuring prism, a thin film of liquid spreads between them. Light reflected by the mirror enters the lower prism and passes into the upper prism. The deviation of the ray of light depends upon the angle of incidence. At an angle near 90° to its surface, the rays will deviate the least on entering the prism and rays entering the measuring prism at angles less than 90° bent more and form the edge of the field. The line of demarcation between the dark and light fields will be coloured and cannot be visualized because when white light is used it will be refracted to a different extent by the measuring prism and the liquid. Light of different wavelengths is dispersed by the liquid, measuring prism, and by the Amici prism A_1. As different liquids produce different dispersion, therefore the second Amici prism A_2 is so adjusted that its dispersion is equal and opposite to that produced by the liquid in measuring prism and A_1. The Amici prism can be adjusted by turning the Milled head until the colour fringe disappears and a sharp light-dark boundary is observed by the eye-piece (Completion of compensation).

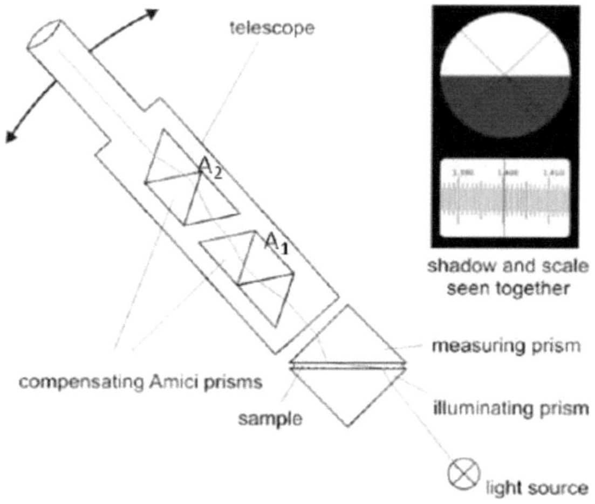

Fig. (7.2). Schematic representation for Abbe Refractometer.

Working Principle of Abbe Refractometer

The instrument is kept on a working table in such a manner that the light from the source is reflected by the mirror. The mirror is adjusted to get the maximum illumination. The prism box is opened using a clamp. The surfaces of the two prisms are cleaned with acetone and dried with soft tissue paper. The polished upper surface should never be scratched. Water is circulated at room temperature through the jacket surrounding the prism, alternatively. Water can be circulated at any temperature with the help of a thermostat.

The refractometer is tested for its correctness. For this purpose, one or two drops of distilled water are placed on the lower surface and the prism box is closed by employing a clamp. The mirror G is rotated in such a way that it reflects light on the prisms. The arc is moved till the borderline appears in the field. The cross wires of the telescope are focused by sliding the eyepiece up and down. The compensator is adjusted with the help of a screw till the coloured line disappears and the borderline coincides exactly with the interaction of the cross wires. The reading is recorded from the scale. The process is repeated 3 to 4 times and the mean of all the values is considered as the refractive index of water. This value should be identical to the standard refractive index of water at room temperature. The prism box is opened and the polished glass surface is cleaned with acetone or ethanol. Now 2-3 drops of experimental liquid are placed on the ground surface of the lower prism. The prism box is closed and the value of the refractive index is recorded as described earlier. Volatile liquids are added through a groove provided in the prism box.

I. TO FIND OUT THE REFRACTIVE INDEX OF THE GIVEN LIQUID AND ALSO FIND ITS MOLAR REFRACTIVITY

Chemicals and Apparatus

Glycerin, density bottle, Abbe refractometer, ordinary light lamp, thermometer.

Theory

The molar refractivity of the liquid is given by:

$$R_M = \frac{M}{d} \times \frac{n^2-1}{n^2+2} \tag{7.5}$$

Where R_M is molar refractivity, M represents the molar weight of the medium, d is density and n represents refractive index.

Procedure

The method consists of two steps:

1. The density of the liquid is determined by using a density bottle.

2. The refractive index of the given liquid is determined using an Abbe refractometer. For description, see the introduction.

Observations

Room temperature = ... t °C

Density of liquid

W_1 = Mass of empty density bottle

W_2 = Mass of empty density bottle + water

W_3 = Mass of empty density bottle + liquid

Therefore, $(W_2 - W_1) = W$= Mass of water at t°C

$(W_3 - W_1) = W'$= Mass of liquid at t°C

The density of water at t°C = D g/cc

Thus, Density of liquid (d) $= \frac{W'}{W} \times D$ g/cc

Refractive Index of the Liquid

Refractive index of the liquid is observed using Abbe's refractometer as described earlier (Table **7.1**).

Table 7.1. Refractive index of the liquid.

Refractive index	Mean

Calculation

The density (d) of the liquid = $\frac{W'}{W} \times D$ g/cc

Knowing all the values we can calculate the value of molar refractivity R_M of the given liquid from Equation (7.5).

Results

The refractive index and molar refractivity of the liquid are .... and ...... respectively.

Precautions

Do not let the polished surface of the prisms of the refractometer get scratched.

II. TO FIND OUT MOLAR REFRACTIVITIES OF THREE LIQUIDS A, B AND C. TO CALCULATE THE COMPOSITION OF LIQUID C, WHICH IS A MIXTURE OF LIQUIDS A AND B

Chemicals and Apparatus

Two liquids say glycerine (A) and water (B), Abbe refractometer, density bottle, ordinary light lamp, and thermometer.

Theory

The refractive indexes of solutions depend on their composition. The percent composition of liquid C can be determined using the formula:

$$100\,(R_M)_c = x(R_M)_A + (100 - x)(R_M)_B \qquad \textbf{(7.6)}$$

Here x is the percentage of A in the mixture C. the value of x can also be determined graphically by plotting a curve between the concentration of A (by weight) *vs.* molecular refractivity. So, the composition of C can be determined from the graph by interpolation corresponding to its refractive index.

Procedure

The refractive index of liquids A, B, and mixture C are determined using the Abbe refractometer (as described earlier in this chapter) and recorded in Table **7.2**. The densities of the three liquids are determined using density bottle as usual.

A number of mixtures containing different concentrations of A and B are prepared to perform the experiment graphically. For example, by dissolving 90%, 80%, 70%, 60%..., 10% of A in water by weight. The refractive index and density of all the compositions are measured by refractometer and density bottle, respectively. The molar refractivities of all the solutions are calculated (Table **7.3**) and plotted as ordinate while the percent composition of A is plotted as abscissa. The molar refractivity of the given mixture C is also calculated after determining its refractive index and density.

Using the calibration curve, we can find the composition of C.

Observations

Room temperature = ...°C

Density of Liquid

W_1 = Mass of empty density bottle.

W_2 = Mass of empty density bottle + water.

W_3 = Mass of empty density bottle + liquid A.

W_4 = Mass of empty density bottle + liquid B.

W_5 = Mass of empty density bottle + liquid C.

Therefore, $(W_2 - W_1) = W$= Mass of water at t°C.

$(W_3 - W_1) = W'$= Mass of liquid at t°C.

The density of water at t°C = D g/cc.

Thus, Density of liquid (d) = $\frac{W'}{W} \times D$ g/cc

Table 7.2. Refractive index measurements.

Liquid	Refractive Index	Mean
A		

B		
C		

Calculations

Table 7.3. Calculation of Molar Refractivity

Liquid	Density (d)	Molecular Weight (M)	Molar Refractivity $R_M = \frac{M}{d} \times \frac{n^2 - 1}{n^2 + 2}$
A	$\frac{w_3 - w_1}{w_2 - w_1}$		
B	$\frac{w_4 - w_1}{w_2 - w_1}$		
C	$\frac{w_5 - w_1}{w_2 - w_1}$		

Knowing the values of molar refractivity of A, B and C we can calculate the percentage composition, x of A in the mixture using Equation (7.6).

Result

The composition of the mixture is ......% A and .... % B. Molar refractivities of the liquids A, B and C are....., ... and .... Respectively.

III. TO CALCULATE ATOMIC REFRACTIVITIES OF CARBON, HYDROGEN AND OXYGEN BY TAKING METHYL ACETATE, ETHYL ACETATE AND n-HEXANE

Chemicals, and Apparatus

Methyl acetate, ethyl acetate, and n-hexane, Abbe refractometer, ordinary light lamp, density bottle.

Theory

Molar refractivity is an additive and constituted property. Hence the molar refractivity of a molecule is the sum of refractivities of the atoms constituting the molecule. Hence,

(a) R_M of CH_2 group = R_M of methyl acetate - R_M of ethyl acetate. **(7.7)**

(b) R_M of n-hexane – 6 × R_M of CH_2 group = 2 × R_M of H atom. **(7.8)**

(c) R_M of CH_2 group – 6 × R_M of H atom = R_M of C atom. **(7.9)**

(d) R_M of methyl acetate – 3 × R_M of C atom - 6 × R_M of H atom = 2 × R_M of O atom (7.10).

Procedure

The refractive index and density of methyl acetate, ethyl acetate, and n-hexane are determined with the help of Abbe refractometer and density bottle (Table **7.4**) as described in the previous experiment.

Observation

Temperature = ……°C

Table 7.4. Table for refractive index, density, and molar refractivity.

Liquid	Refractive Index	Density	Molecular Weight	Molar Refractivity $R_M = \frac{M}{d} \times \frac{n^2 - 1}{n^2 + 2}$
Methyl acetate				
Ethyl acetate				
n-hexane				

Calculations

After determining molar refractivities of methyl acetate, ethyl acetate, and n-hexane we can determine the atomic refractivities of H, C, and O atoms using Equations (7.7) –(7.9).

Result

The atomic refractivities of C, H, and O are …., …... and ……., respectively.

Precautions

Do not let the polished surface of the prisms of the refractometer get scratched.

POLARIMETRY

History

Optical activity was first observed by French physicist Jean Baptiste Biot. He concluded that the change in direction of plane-polarized light when it passes through certain substances is a rotation of light. His work was supported by the experimentation of Louis Pasteur. He observed the existence of two crystals that were mirror images of tartaric acid. Through meticulous experimentation, he found that one set of molecules rotated polarized light clockwise, while the other rotated counterclockwise to the same extent. He observed that racemic mixture did not rotate light, because the optical activity of one molecule cancels the effects of another molecule. The existence of chiral molecules was first shown by Louis Pasteur.

Introduction

The identification of some substance is facilitated by eliminating the vibrations of light radiation. The process is referred to as polarization. When one vibration is eliminated it is called plane-polarized. That means the light is directed along a particular plane. A large number of substances characteristically rotate the plane of polarisation. These substances are referred to as optically active *e.g.*, tourmaline crystal which demonstrates that only transverse vibration of radiation is passed through it while perpendicular vibration is cut off. This activeness of some substances is due to their structure and due to asymmetric nature. Those molecules which show such activity, are called **optically active material**. Chiral molecules can interact with the electric vector of plane-polarized light, which leads to a change in the direction of the vector. This leads to optical activity. For achiral molecules such component of interaction is absent. So achiral molecules do not show optical activity.

Many substances characteristically rotate the plane of polarization. Optical rotatory power has its origin in structural asymmetry.

Polarization

Specific Rotation

At a given temperature, t and for a given wavelength 'λ', specific rotation is defined as the rotation (in degrees) produced by a path of one decimeter in a substance of unit density. If 'θ' is the rotation produced by 'l' decimeter length of a solution of density d g /cc, then the specific rotation S at a given temperature t and for a given wavelength λ is:

$$S_{\lambda}^{t} = \frac{\theta}{l \times d} \tag{7.11}$$

$$\text{Unit of specific rotation} = \frac{\text{Rotation in Degrees}}{\text{Length in decimeter} \times \text{concentration in g/cc}} \tag{7.12}$$

Molecular Rotation

For comparison of rotating power of different substances molecular rotation M_{λ}^{t} functions are defined by the equation:

$$M_{\lambda}^{t} = \frac{M \times S_{\lambda}^{t}}{100} = \frac{M \times R}{l \times c} \tag{7.13}$$

Where R is Angle of Rotation and R is defined as:

$$R = S_{\lambda}^{t} l \frac{m}{V} \tag{7.14}$$

Here l is the length of the column in decimeters through which the light passes and m is weight in gram (g) of optically active substance dissolved in a volume of v ml.

Laurent's Polarimeter

Half shade device: To overcome the ability of the eye to judge the exact position of extinction, when the two Nicols are adjusted in crossed position, a half shade device is designed. It consists of two semi-circular plates, one made of glass and the other made of quartz. Both the halves are cemented together and the quartz plate cuts the parallel component to the optic axis.

Lippich Polarimeter

Laurent's half shade device is suitable for only one particular wavelength for which a quartz half-wave plate is made. To overcome this difficulty Laurent's polarimeter and Biquartz polarimeters are designed. In Lippich Polarimeter, a Lippich prism is used instead of quartz half shade. Lippich prism is a small Nicol prism.

In Biquartz polarimeters, two quartz semicircular plates are cemented together out of which, one is left-handed and the other one is right-handed.

Applications

Sugar Industry, Food product Industry, Pharma Industry, and Chemical Industry /Laboratories.

Basics – Plane Waves

The simplest manifestation of polarization to visualize is that of a plane wave (Fig. **7.3**). A plane wave is a good approximation of most light waves (a plane wave is a wave with infinitely long and wide wavefronts). All electromagnetic waves propagating in free space or a uniform material of an infinite extent have electric and magnetic fields perpendicular to the direction of propagation. Conventionally, when considering polarization, the electric field vector is described and the magnetic field is ignored since it is perpendicular to the electric field and proportional to it. The electric field vector may be arbitrarily divided into two perpendicular components labeled and y (with z indicating the direction of travel). For a simple harmonic wave, where the amplitude of the electric vector varies in a sinusoidal manner, the two components have the same frequency. However, these components have two other defining characteristics that can differ. First, the two components may not have the same amplitude. Second, the two components may not have the same phase, that is they may not reach their maxima and minima at the same time. The shape traced out in a fixed plane by the electric vector as such a plane wave passes over it is a description of the polarization state.

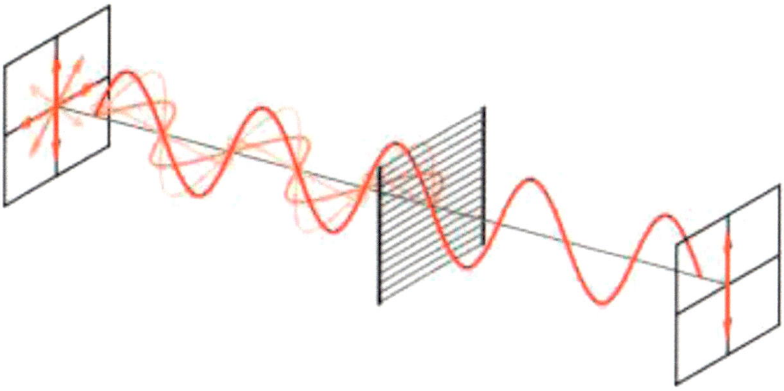

Fig. (7.3). Propagation of plane-polarized light.

In practice, some light is lost in the polarizer and the actual transmission of unpolarized light will be somewhat lower than this, around 38% for Polaroid-type polarizers but considerably higher (>49.9%) for some birefringent prism types.

If two polarizers are placed one after another (the second polarizer is generally called an analyzer), the mutual angle between their polarizing axes gives the value of θ in Malus' law. If the two axes are orthogonal, the polarizers are crossed and in theory, no light is transmitted, though again practically speaking no polarizer is perfect and the transmission is not exactly zero (for example, crossed polaroid sheets appear slightly blue). If a transparent object is placed between the crossed polarizers, any polarization effect present in the sample (such as birefringence) will be shown as an increase in transmission. The ratio of the transmission of the unwanted component to the wanted component is called the extinction ratio and varies from around 1:500 for polaroid to about $1:10^6$ for Glan-Taylor prism polarizers.

Optical rotation depends mainly on (a) wavelength (λ), (b) path length (ι), (c) temperature (t), (d) density (p) or concentration (c) of the solution. The rotation is measured by a term called specific rotary power (α):

$$\alpha = 100\ \theta / l\ wp \qquad \textbf{(7.15)}$$

Where θ, l, w & p are angular rotation of sample in degrees, length in decimeters, the weight of solute in g of the solution, and density of the solution, respectively.

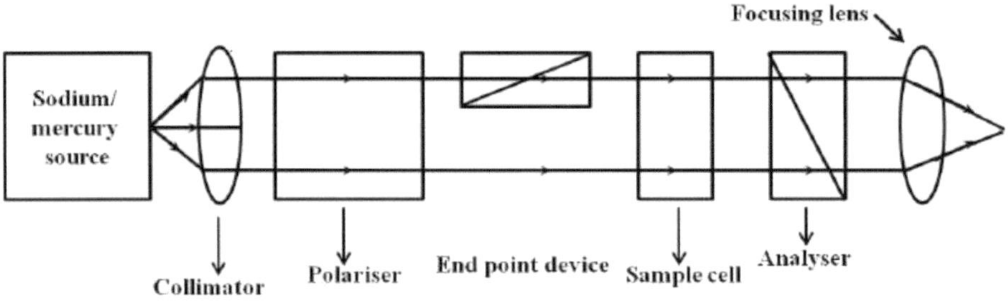

Fig. (7.4). Schematic diagram of a polarimeter.

The simple polarimeter (Fig. **7.4**) consists of the following modules:

A. Source: Since the specific rotary power has a strong wavelength dependence and varies considerably with each wavelength, the source employed is usually monochromatic. The cheapest source is sodium light or mercury lamp with a green filter attached to it.

B. Collimator: The beam from Na or Hg lamp is allowed to fall on the collimator. The collimator is an achromatic lens, which focuses the beam on the polariser.

C. Polariser: The polariser is a crystal of calcite. This crystal rotates the plane of vibration of radiation. Thus, this gives a linearly polarised beam that is now concentrated on an endpoint device.

D. Endpoint device: Since the eye has a poor sensing efficiency for measuring the plane of rotation, it is necessary to alter the plane of polarisation of 1/2 or 1/3 of the beam energising from the polariser. In effect light after passing through the endpoint device splits into two parts one having more intensity and the other less. These two beams now fall on the sample which further introduces characteristic rotation.

E. Sample Cell: After passing through the sample, the two light beams pass through an analyser made of calcite crystal, which further rotates the plane of rotation. When the polariser and analyser are at right angles, the intensity is minimum. When the analyser is further rotated a half-shadow, appearance is observed. Note the angle at this stage. The difference in two angular settings gives θ. Thus, knowing the value of θ, l, w & p, one can find 'α'.

F. Analyser: Analyser is made up of calcite crystal. In a colorimeter, a beam of light with a specific wavelength is passed through a solution *via* a series of lenses, which navigate the colored light to the measuring device. This analyzes the color compared to an existing standard. A microprocessor then calculates the absorbance or percent transmittance. If the concentration of the solution is greater, more light will be absorbed, which can be identified by measuring the difference between the amount of light at its origin and that after passing the solution.

I. TO FIND OUT THE SPECIFIC ROTATION AND MOLECULAR ROTATION OF AN OPTICALLY ACTIVE SUBSTANCE POLARIMETRICALLY AND ALSO FIND CONCENTRATION OF UNKNOWN SOLUTION

Chemical and Apparatus Required

Optically active substance, say cane sugar, polarimeter, chemical balance, beaker, soft tissue paper.

Theory

The relation between the specific and molecular rotation of a substance can be expressed by the following equations:

$$S_{\lambda}^{t} = \frac{100\,R}{lc} \tag{7.16}$$

$$M_{\lambda}^{t} = \frac{MR}{lc} \tag{7.17}$$

Here all symbols carry usual significance. The concentration of the unknown solution can be determined by means of a calibration curve, plotted between concentration and rotation. The specific rotation has been observed to vary with the concentration of the solution, that's why another concentration-independent function known as intrinsic rotation, has been given. Intrinsic rotation of any solute in a given solvent for a given wavelength of light and given temperature, is the limiting value of a specific rotation when concentration tends to zero. This can be determined by extrapolating the S *vs.* c graph for c = 0. $S_{\lambda}^{t} = \{S\} + bc$. $\{S\}$ is intrinsic rotation.

Procedure

1) Weigh 5 g of cane sugar and dissolve it in distilled water in a 100 ml measuring flask and make the volume up to the mark by adding distilled water. Ensure the solution contains no turbidity. If turbidity appears, filter the solution. If turbidity remains, discard and prepare a fresh solution. Rinse the polarimeter tube with the same solution and fill up it. If some bubble appears, pass it to the bulb of the polarimeter.

2) Record the zero reading by filling the polarimeter tube with distilled water. Wipe the polarimeter tube and lenses to remove the solution and other impurities including fingerprints using soft tissue paper. If on placing the tube filled with cane sugar solution, the analyser turns dark on the right, the substance is dextro-rotatory and if it is dark on the left, it is laevorotatory. To record the reading, rotate the adjusting screw to obtain a position when two halves are equally illuminated. Record 2 or 3 more readings and take the most precise value.

3) Prepare 1%, 2%, 3%, and 4% solutions of cane sugar from the stock solution and determine the angle of rotation for each concentration (Table **7.5**). Keep in mind that before filling the polarimeter tube with a solution of any specific concentration, it must be rinsed twice with it. Similarly, determine the angle of rotation for an unknown solution.

Observation

Room temperature = ……°C.

Zero reading with distilled water = ...r°.

Length of polarimeter tube = ….dm.

Table 7.5. Angle of rotation of cane sugar at various concentration.

S. No.	Concentration of Cane Sugar Solution (% w/v)	Reading in Degree r_1°	Angle of Rotation R° ($R = r_1 - r$)
1	1	------	------
2	2	------	------

3	3	------	------
4	4	------	------
5	5	------	------
6	Unknown	------	------

Calculations

The specific and molecular rotation of cane sugar can be calculated from any solution by using Equations (7.16) and (7.17).

A graph is plotted between concentrations and the angle of rotation. A straight line is obtained. With help of this graph, the concentration of an unknown solution can be determined.

Intrinsic rotation can be determined by plotting as S_{λ}^{t} *vs*. Concentration. Extrapolate it to c = 0, *i.e.*, intercept on the y-axis.

Result

The specific rotation of cane sugar at .... °C = ... °.

Molecular rotation of cane sugar at .... °C = ... °.

The concentration of unknown solution = ... %.

Intrinsic rotation of cane sugar = ... °.

Precautions

There should be no air bubble in the polarimeter tube.

The metal caps should be screwed lightly to avoid a strain on the glass tube.

The glass plates must be clean and the outer surface must be dry.

The solution should be absolutely clear.

Always use freshly prepared cane sugar.

II. TO FIND OUT THE PERCENTAGE OF D-SUGAR AND D-TARTARIC ACID IN A GIVEN SOLUTION POLARIMETRICALLY

Chemicals and Apparatus Required

Polarimeter, stopwatch, *d*-sugar, and *d*-tartaric acid, beakers, soft tissue paper.

Theory

When two solutes A and B do not interact and are optically active, the specific rotations due to the two solutes are additive. Hence, the rotation of the mixture (M) solution is:

$$[\alpha]_M = C_A[\alpha]_A + C_B[\alpha]_B = C_A[\alpha]_A + (1-C_A)\,[\alpha]_A \qquad \textbf{(7.18)}$$

Here, C stands for the mole fractions of the solute, and α stands for specific rotation of the respective solutes. Thus,

$$[\alpha]_M = C_A[\alpha]_A + [\alpha]_B - C_A[\alpha]_B$$

$$= C_A([\alpha]_A - [\alpha]_B) + [\alpha]_B \qquad \textbf{(7.19)}$$

$$C_A = \frac{([\alpha]_M - [\alpha]_B)}{([\alpha]_A - [\alpha]_B)} \qquad \textbf{(7.20)}$$

In a similar manner, we can calculate C_B as:

$$C_B = \frac{([\alpha]_M - [\alpha]_A)}{([\alpha]_A - [\alpha]_B)} \qquad \textbf{(7.21)}$$

Thus, the amount of each solute can be calculated in the mixture by determining the specific rotations of solutes A, B, and mixture.

Procedure

1) Prepare 5% solutions of *d*-sugar and *d*-tartaric acid.

2) The specific rotation of the standard solutions, as well as that of an unknown mixture (Table **7.6**), is determined as described in experiment I with polarimeter (To find out the specific rotation and molecular rotation of an optically active substance polarimetrically and also find the concentration of unknown solution).

Observations

Room temperature = …. °C.

Zero reading with distilled water = …..r°.

Length of polarimeter tube = …... dm.

Table 7.6. Angle of rotation at various concentrations.

S. No.	Concentration of Solution (% w/v)	Reading in Degree r_1°	Angle of Rotation R° ($R=r_1-r$)
1	5% *d*-sugar	------	------
2	5% *d*-tartaric acid	------	------
3	Unknown	------	------

Calculations

The values of $[\alpha]_A$, $[\alpha]_B$, and $[\alpha]_M$ are calculated as described in the first experiment. From Equations (7.20) and (7.21), the values of C_A and C_B can be calculated.

Results

The composition of mixture is = ……% *d*-sugar and …… % *d*-tartaric acid.

Precautions

Prepare the standard solutions of *d*-sugar and *d*-tartaric acid accurately.

III. TO FIND OUT ORDER OF A REACTION AND VELOCITY CONSTANT FOR THE INVERSION OF CANE SUGAR BY ACID

Chemicals and Materials Required

Sugar, 100 ml (N/2) hydrochloric acid (HCl), Polarimeter, beaker, stopwatch.

Theory

Cane sugar is a disaccharide, which hydrolyse in presence of an acid, giving glucose and fructose as products:

$$\underset{\text{Cane sugar}}{C_{12}H_{22}O_{11}} \xrightarrow{H^+} \underset{\text{Glucose}}{C_6H_{12}O_6} + \underset{\text{Fructose}}{C_6H_{12}O_6}$$

The solution of cane sugar is dextrorotatory. It undergoes hydrolysis in presence of acid to give glucose (dextrorotatory) and fructose (leavo rotatory) in equal amounts. The leavo rotation of fructose is -92° and the dextrorotation of glucose is +52.5°. Thus, the mixture is leavo rotatory. This is the reason why the process of hydrolysis of cane sugar is known as inversion. The advancement of the reaction can be policed by recording the change in angle of rotation by means of a polarimeter. The changes in rotation in a time interval, say t, give a measure of the amount of cane sugar hydrolysed (x). The change in rotation produced on completion of reaction gives a measure of the initial concentration of cane sugar (a).

As only the concentration of cane sugar changes during the reaction (pseudo unimolecular reaction), the reaction is of the first order. Water is present in large excess, so the concentration of water does not change significantly. For first-order reaction the rate expression can be given as:

$$k = \frac{1}{t} \ln\left(\frac{a}{a-x}\right) = \frac{2.303}{t} \log\left(\frac{a}{a-x}\right) \quad \textbf{(7.22)}$$

Here k is rate constant. For inversion of cane sugar, $a = r_0 - r_{inf}$ and $x = r_0 - r_t$, where, r_0, r_t, and r_{inf} is the rotation at the beginning of the reaction, after time t and on completion reaction, respectively. So,

$$k = \frac{2.303}{t} \log_{10}\left(\frac{r_0 - r_{inf}}{r_t - r_{inf}}\right) \quad \textbf{(7.23)}$$

Procedure

1) Prepare 100 ml of 5% cane sugar solution.

2) Take a blank reading with distilled water and then with the cane sugar solution. Now mix 100 ml of this solution to 100 ml of 0.5 (N) HCl in a reaction vessel. Immediately fill this solution into a polarimeter tube and start taking readings. As soon as the first reading is taken start the stopwatch. This reading corresponds to the rotation value (r_0) at the start of the reaction.

3) Record the readings after time intervals of 5, 10, 20, 30, 40, and 50 min. Record the final reading after keeping the reaction mixture for 24 hours or keeping the reaction mixture at 60°C in a water bath for 30 min. This value of angle of rotation

corresponds to r_{inf}, *i.e.*, heating after the reaction. Record the observations in Table **7.7**.

Observations

Room temperature = …. °C.

Zero reading with distilled water = …..r°.

Table 7.7. Kinetics of inversion of cane sugar.

S. No.	Time (min)	Reading (r_1°)	Angle of Rotation (r_1- r°)	Rate Constant k (min^{-1})	Mean Rate Constant (min^{-1})
1	0	…..	…..	…..	-
2	5	…..	…..	…..	
3	10	…..	…..	…..	
4	20	…..	…..	…..	
5	30	…..	…..	…..	
6	40	…..	…..	…..	
7	50	…..	…..	…..	
8	infinity	…..	…..	…..	

Now plot t *vs.* $\log_{10}\left(\frac{r_0 - r_{inf}}{r_t - r_{inf}}\right)$ will give a straight line passing through the origin and the slope of the straight line is k (Fig. **7.5**). Thus, slope k can also be calculated.

The value of k calculated for every time interval comes out is constant, showing the reaction to be first order.

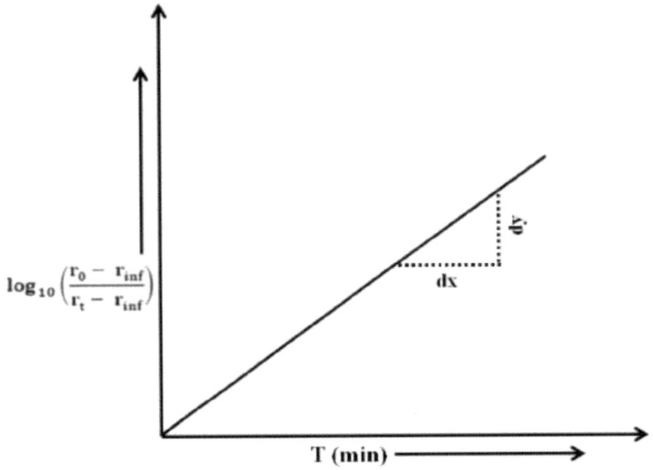

Fig. (7.5). Graphical representation for inversion of sucrose.

Results

The reaction is first order with the value of the rate constant at .... °C =.... Min^{-1} (From table).

The reaction is first order with the value of the rate constant at .... °C =.... Min^{-1} (From plot).

The values of k from the calculation and graph should be in good agreement.

Precautions

Same as in the previous experiment.

COLORIMETRY AND SPECTROPHOTOMETRY

Introduction

A colorimeter is an optical instrument that measures the colour concentration of a substance in a solution.

Principle

This instrument is based on Lambert-Beer's law. According to Lambert's law, when light radiation travels through an absorbing medium, the rate of decrease of intensity of radiation with the distance (l) traveled is proportional to the intensity of incident radiation. Mathematically:

$$-\frac{dI}{dl} = k'l$$

$$\text{or, } -\frac{dI}{I} = k'dl \qquad \textbf{(7.24)}$$

Integrating Equation (7.24) we get,

$$\int_{I_0}^{I_1} \frac{dI}{I} = -k' \int_0^1 dl$$

$$\text{or, } \ln\left(\frac{I_1}{I_0}\right) = -k'l$$

$$\text{or, } \log\left(\frac{I_1}{I_0}\right) = -k'l/2.303 \qquad \textbf{(7.25)}$$

On the other hand, Beer's law states that the intensity of absorption of radiation also depends on the concentration c of light-absorbing species.

Thus, a combination of both, which is known as Lambert-Beer's law can be expressed as:

$$\log\left(\frac{I_1}{I_0}\right) = -\varepsilon cl$$

$$\text{or, } A = -\varepsilon cl \qquad \textbf{(7.26)}$$

Where A is absorbance and A= $\log\left(\frac{I_1}{I_0}\right)$; ε, c and l are molar absorption coefficient, concentration, and path length.

The Instrument

The principle of colorimetry is based on the following flow chart as shown in Fig **(7.6)**.

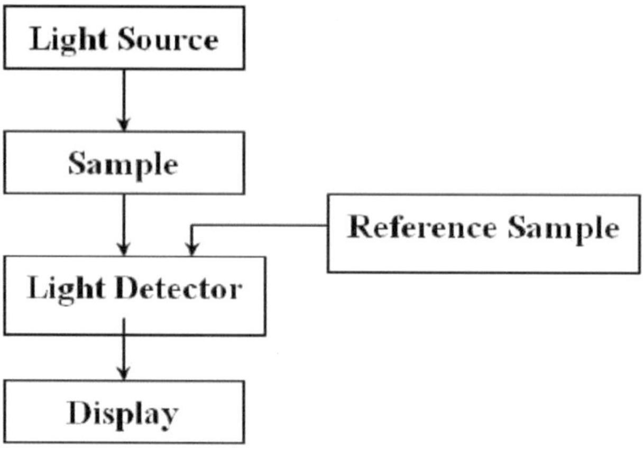

Fig. (7.6). Flowchart to explain the principle of colorimeter.

The construction of the instrument is shown in Fig. (**7.7**). A light source is a low voltage lamp (6.3 V). The light passes through an optical filter. This light is focused on an optical system. The light falls on a cuvette containing liquid.

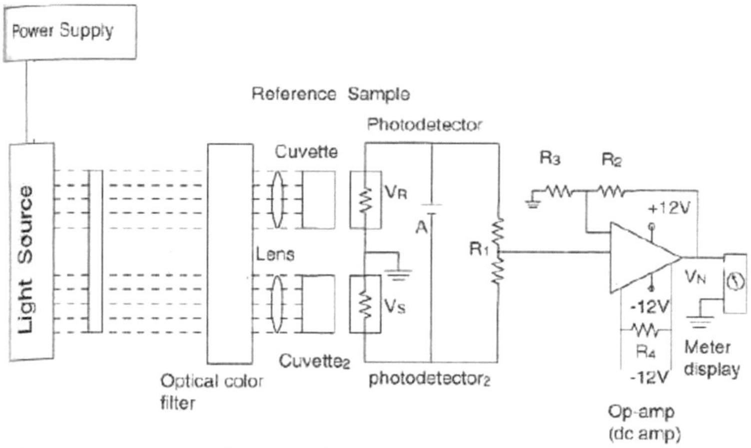

Fig. (7.7). Schematic representation of Basic colorimeter.

The light is further absorbed by the liquid. Then the light goes to the photodetector circuit. The amplification of photodetector current is made using an amplifier. This drives the meter. Two cuvettes are used simultaneously. A comparison of output and photodetector is made using an operational amplifier. The instrument needs calibration. It is done in the following way.

First, the operational amplifier error is removed by balancing the potentiometer. For this, both inputs are grounded. Then, a reference liquid-like distilled water is kept in both cuvettes. The output of the indicator is adjusted to zero. Both cuvettes contain the same fluid hence the differential output is zero. Next, the reference liquid is retained in cuvette 1, and cuvette 2 is replaced by the solution/liquid to be investigated. Now the differential reading given by the meter is an indicator of liquid absorbance of light.

Construction

To make the instrument ready to use, set zero and set 100% controls are available on the panel. A digital indication of percentage concentration is provided on screen. ON-OFF controls are provided. The zero adjust control is used when reference liquid is selected. The 100% adjust control, is used when standard 100% absorbing liquid is selected. Filters are mounted on a turret. By rotating this, various filters can be selected. The instrument work on a 230 V supply. Following filters (Table **7.8**) are available:

Table 7.8. Filters and wavelength range.

Filter No.	Range (nm)
1	420-560
2	480-590
3	540-610
4	570-700
5	600 nm and above

Maintenance

The instruments maintenance includes regular calibration. The burnt lamp needs replacement. A cooling fan is provided in some instruments, which needs replacement when damaged. The electronic circuit has the least problems. Faulty photocell, dirty reflector can cause problems.

SPECTROPHOTOMETER

Introduction

The spectrophotometer is a routinely used instrument in scientific research. Spectrophotometry is the quantitative measurement of how much a chemical substance absorbs light by passing a beam of light through the sample using a spectrophotometer.

Principle

The absorbance is measured at various wavelengths. Every substance has a particular absorption capacity to various wavelengths. This can be used to detect a component of unknown material in a certain liquid. The block schematic of the principle is explained in Fig. (**7.8**).

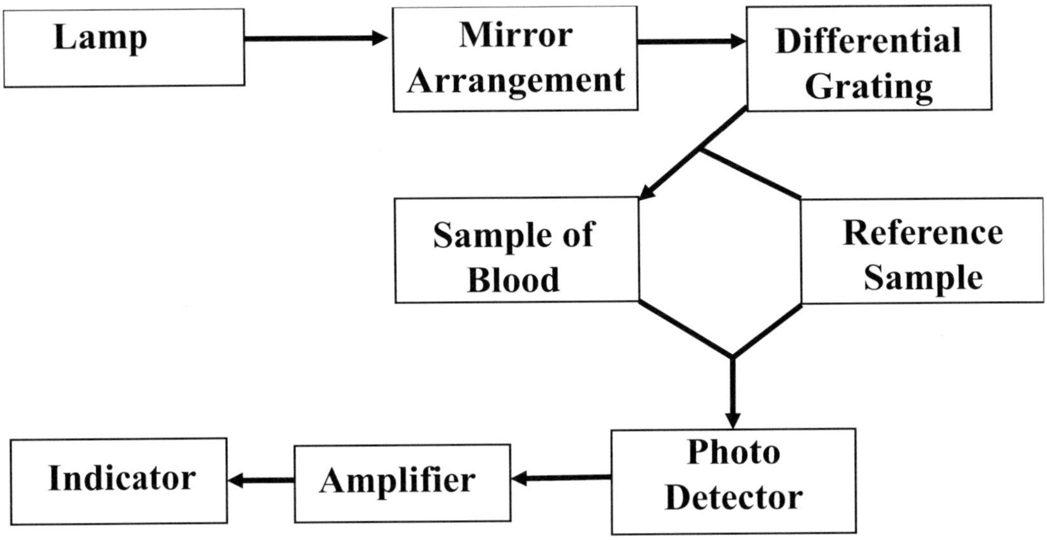

Fig. (7.8). Flowchart to explain the principle of Spectrophotometer.

Instrumentation

A lamp is provided as a source of light. The instrument block diagram is shown in Fig. (**7.9**). The light passes through slit S and is then reflected by a mirror. The light is diffracted using a diffraction grating. The construction of grating is very important. As the wheel outside the meter is rotated the angle of grating changes. This change of angle changes the travel of light further.

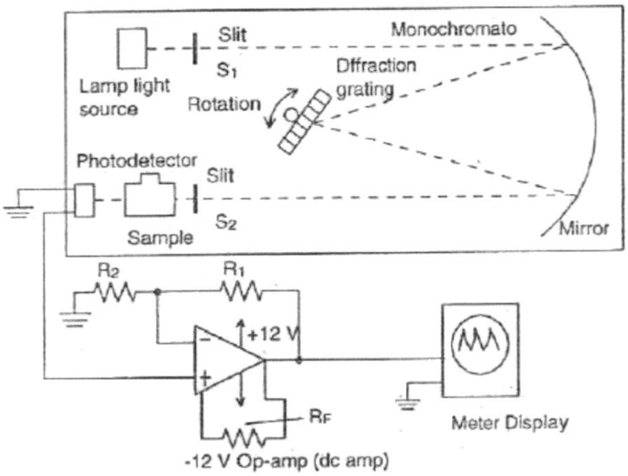

Fig. (7.9). Spectrophotometer simplified schematic diagram.

The light consists of numbers and colors. Violet to red colors is observed in the spectrum. The frequency of the visible spectrum varies from 10^{-6} to 10^{-8} meters. The spectrophotometer uses a fused silica lamp (200-340 nm) or a tungsten halogen lamp (340-900 nm). Lamp-generated light is passed through the selector mechanism. The lamp light is passed through a slit for band selection. The grating is used for frequency selection. The reflection mirrors pass light through the sample to a photomultiplier tube. The electrical output of the photomultiplier tube is amplified and given to the indicator. A Diagram of a typical spectrophotometer is given in Fig. (**7.10**).

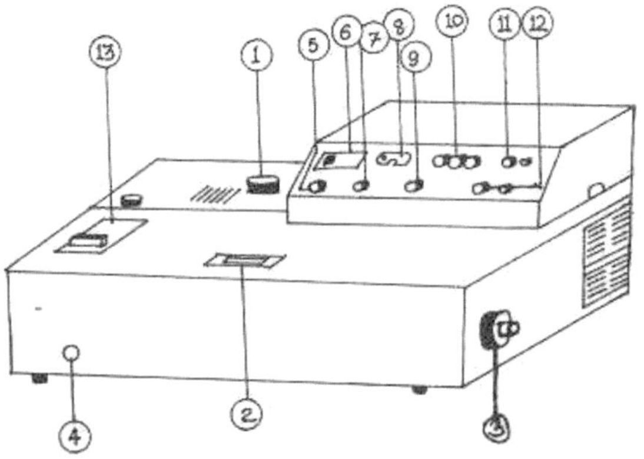

Fig. (7.10). A typical spectrophotometer. In the figure 1=>Sourceselector, 2=> Wavelength window, 3=> Wave length adjust, 4=> Sample position control, 5=> Shutter, 6 => Set zero, 7=> Digital display, 8=> AMP Gain (Fine), 9=> AMP gain (COVERSE), 10 => Set concentration, 11=> Source ON/OFF, 12 => Tungsten Lamp, 13 => Lamp.

Operations

Steps for operation are:

1. For the ultra-violet region (200-340 nm) deuterium lamp is selected. Filters should be selected from the following wavelength (nm):

200	-	360
360	-	500
500	-	560
560	-	630
630	-	900

The desired wavelength should be selected from the indicator provided and wheel operation.

2. Solution of reference liquid *i.e.*, distilled water should be filled in the cuvette.

3. Adjustment of zero is done on indicator with zero adjustment.

4. 100% adjustment is made by pulling out the shutter (dark).

5. Test solution is kept in the cuvette and the display will indicate absorbance.

Construction

The controls for zero adjustment, 100% adjustment, sensitivity, and digital display are provided on the front panel. The cuvettes can be pushed back or forward by a knob in the front. Also, the wheel at the right end can be rotated to select a frequency. An indicator is available for frequency selection.

Maintenance

The ultraviolet lamp used in the spectrophotometer needs care and replacement. Mechanical shock can damage the instrument; hence care must be taken for the operation. Cuvettes must be cleaned periodically. The front facia should be used for calibration. Wrong mechanical assemblies of mirrors and grating may create a malfunction.

A spectrophotometer is an instrument used for the measurement of light absorbed by liquid at selected wavelengths. This instrument is available in sophisticated form as (Ultraviolet) Spectrophotometer and has ultraviolet and tungsten lamps both as light sources. Anyone can be selected for measurement. This instrument is used for the analysis of compounds absorbing UV or visible radiation.

I. TO OBTAIN THE CALIBRATION CURVE FOR A GIVEN COMPOUND AND (A) VERIFY THE LAMBERT-BEER'S LAW AND (B) DETERMINE CONCENTRATION OF UNKNOWN SOLUTION

Chemicals and Apparatus Required

0.002 (M) potassium permanganate ($KMnO_4$) or potassium dichromate ($K_2Cr_2O_7$) solution, spectrophotometer/colorimeter (with proper filter), burettes, pipettes, cuvette.

Theory

The principle of the experiment, *i.e.*, Lambert-Beer's (LB) law is discussed in the first part of the article of the section. The instrumentation of both spectrophotometer and colorimeter is also discussed earlier.

Procedure

1) Switch on the supplied instrument and wait for about 15 min for the instrument to get stabilised. By that time prepare a standard stock solution of the given compound in distilled water. Prepare solutions of different concentrations by mixing 9, 8, ... 3, 2, 1 ml of stock solution and 1, 2, ..., 8, 9 ml of distilled water, respectively.

2) If a spectrophotometer is used, then determine the wavelength of maximum absorption (λ_{max}) for the given compound. Set the wavelength to λ_{max} and adjust the percentage transmission for distilled water in the cuvette to 100. Else, if a colorimeter is used, then set absorbance to zero.

3) Fill another cuvette with one of the test solutions and place it in the holder. Record the absorbance in Table **7.9**. Similarly, record the absorbance for all the concentrations, prepared earlier.

Observation Table

Temperature = ...°C; λ_{max} = ...nm.

Table 7.9. Concentration *vs.* absorbance table.

Solution No.	Concentration of Given Compound (M)	Absorbance
1	0	
2	0.0002	
3	0.0004	
4	0.0006	
5	0.0008	
6	0.0010	
7	0.0012	
8	0.0014	
9	0.0016	
10	0.0018	
11	Unknown	

Calculations and Results

Plot the absorbance *vs.* concentration for the given compound to obtain a calibration curve (Fig. **7.11**). A linear plot proves the validity of LB law (mentioned earlier). Determine the concentration of an unknown solution of the given compound from the calibration curve, corresponding to its absorbance.

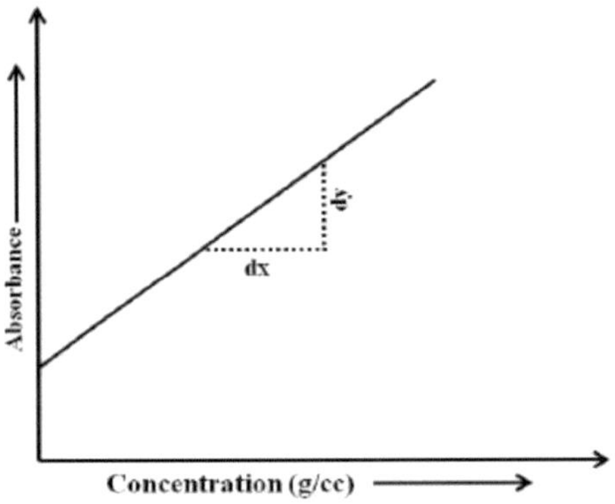

Fig. (7.11). Plot of Concentration *vs.* Absorbance.

II. TO FIND COMPOSITION OF FERRIC ION IN THIOCYNATE COMPLEX BY JOB'S METHOD

Chemical and Apparatus

Ferric nitrate, sodium thiocyanate, nitric acid, hydrochloric acid, spectrophotometer and pipette, conical flask.

Theory

Composition of Complexes

Job's method can be used to determine the formula of a single compound or a complex ion formed by mixing two substances in a solution state. When equimolecular solutions of components are mixed in varying proportions the maximum amount of complex is formed at equilibrium when two components are mixed in the same ratio as required for complex formation. Suppose a complex formed is by the following reaction as:

$$M^{m+} + nL \leftrightharpoons [ML_n]^{(m-n)+}$$

Here M is a metal ion and L is a ligand. The value of n can be determined spectrophotometrically, by measurement of absorbance of a series of mixtures of M and L with varying concentrations. The difference between each observed value of absorbance and that calculated on the basis of no reaction is plotted against the composition of the complex. The value of n can be obtained with the help of molar concentrations of M and L corresponding to minima or maxima of the curve. The reaction is complete if minima or maxima are obtained as a point of intersection of two straight lines. In the case of a reversible reaction, continuous curves are obtained.

Suppose, a part of the solution of M and (1-a) parts of a solution of L (having the same molarity are mixed. If C_M, C_L and $C_{(ML)_n}$ be the equilibrium concentrations of M, L, and ML_n, respectively, then,

$$C_M = Ma - C_{ML_n} \quad \textbf{(7.27)}$$

$$C_L = M(1-a) - nC_{ML_n} \quad \textbf{(7.28)}$$

$$K = \frac{C_{ML_n}}{C_M . C_L^n} K = \frac{C_{ML_n}}{C_M . C_L^n} \quad \textbf{(7.29)}$$

Here, *M* is the molarity of metal ion (M) in the solution. K is stability constant. The reciprocal of K is known as instability constant. For obtaining maxima in the curve of $C_{(ML)_n}$ *vs.* a,

$$\frac{dC_{(ML)n}}{da} = 0 \tag{7.30}$$

On differentiating Equations (7.27), (7.28), and (7.29) and combining them with Equation (7.30):

$$n = \frac{1-a}{a} \tag{7.31}$$

Thus, the value of n can be determined, if the value of a is known for which $C_{(ML)_n}$ is maximum. If the absorbance of the mixture is D and the width of the cuvette, l, then

$$D = I\left(\varepsilon_M C_M + \varepsilon_L C_L + \varepsilon_{ML_n} C_{ML_n}\right) \tag{7.32}$$

If there is no reaction, between M and L, the absorbance D′ of the mixture can be determined by:

$$D' = I\left(\varepsilon_M Ma + \varepsilon_L M(1-a)\right) \tag{7.33}$$

Subtracting Equation (7.32) from Equation (7.33) we have:

$$D - D' = \Delta D = I\left(\varepsilon_M C_M + \varepsilon_L C_L + \varepsilon_{ML_n} C_{ML_n} - \varepsilon_M Ma - \varepsilon_L M(1-a)\right) \tag{7.34}$$

On substituting the values of C_M and C_L from Equation (7.27) and Equation (7.28) in Equation (7.34)

$$\Delta D = IC_{ML_n}\left(\varepsilon_{ML_n} - \varepsilon_M - n\varepsilon_L\right) \tag{7.35}$$

On differentiating Equation (7.34) w.r.t a, we get:

$$\frac{d\Delta D}{da} = I\left(\varepsilon_{ML_n} - \varepsilon_M - n\varepsilon_L\right) \tag{7.36}$$

ΔD will have maximum value, when C_{ML_n} is maximum. In that case $C_{ML_n} > (\varepsilon_M + n\varepsilon_L)$. Similarly, ΔD will have minima, when C_{ML_n} has a minimum value, *i.e.*, $C_{ML_n} < (\varepsilon_M + n\varepsilon_L)$. The value of n can be determined from the maximum and minimum position of the curve, between ΔD and a.

Getting either maxima or minima indicates the formation of only one complex, while two or more maxima or minima indicate the formation of two or more complexes.

The complex between ferric ion and thiocyanate ion is formed as:

$$Fe^{3+} + nCNS^{-} \leftrightharpoons Fe(CNS)_n^{3-n}$$

The value of n can be determined by using Equation (7.31).

Procedure

1) Prepare 4×10^{-3} (M) solution of $Fe(NO_3)_3$ in 0.02 (M) HNO_3 with total ionic strength of 0.04. The role of HNO_3 is the suppression of hydrolysis of $Fe(NO_3)_3$ due to the reduction of pH.

2) Similarly, prepare 4×10^{-3} (M) solution of NaCNS in 0.036 (M) HCl with ionic strength 0.04. HCl is added to make the same total ionic strength in the two solutions.

3) Prepare a series of solutions with varying concentrations by adding 9.5, 9.0, 8.5, ...0.05 ml of $Fe(NO_3)_3$ and 0.05, 1.0, 1.5,....., 9.5 ml of NaCNS, respectively.

4) Now determine the wavelength of maximum absorbance (λ_{max}) by measuring the absorbance of each mixture using water as blank, in the wavelength range 380-540 nm. λ_{max}= 460 nm.

5) Now measure the absorbance of solutions containing pure compounds and the mixture against water at this wavelength. The absorbance of the original solution is treated to be zero. So, ΔD will be the absorbance of the mixture.

Observations and Calculations

The value of absorbance of mixture *vs*. the composition of the mixture is plotted to find the composition corresponding to maxima and hence the value of n can be calculated.

III. TO DETERMINE THE COMPLEX FORMATION BETWEEN FERRIC AND SALICYLIC ACID TO FIND THE FORMULA AND STABILITY CONSTANT OF THE COMPLEX

Chemical and Apparatus Required

0.002 (M) HCl, 0.001 salicylic acid, and Ferric ammonium alum dissolved separately in 0.002 (M) HCl, spectrophotometer, burettes, labeled stopper glass tubes.

Theory

Ferric ion (Fe^{3+}) and salicylic acid form a complex in a solution having a pH range of 2.6-2.8. At this pH phenolic, OH, and COOH groups remain unionised. Suppose the complex formation occurs according to the reaction given below:

$$Fe^{3+} + nC_6H_4(OH)COOH \rightarrow Fe[C_6H_4(OH)COOH]_n^{3+}$$

The stoichiometry of the reaction can be determined by Job's method. When equimolar solutions of the reactant are mixed in varying proportions, the maximum amount of complex is formed at equilibrium, when the proportions of the reactants used correspond to the empirical formula of the complex. If x and (1-x) are respective parts of equimolar solutions of Fe^{3+} and salicylic acid in the mixture corresponding to the maximum amount of the complex formed, then, $n = \frac{1}{x} - 1$. The equilibrium concentration can be followed by measuring the absorbance of the solution because the complex is coloured.

The stability or equilibrium constant for the complex formation is given by

$$K = \frac{[Fe[C_6CH_4(OH)COOH]_n^{3+}]}{[Fe^{3+}][C_6CH_4(OH)COOH]^n} \tag{7.37}$$

The stability of the Fe (III)-salicylic acid complex, in the pH range 2.6-2.8, is required to have maxima in the curve of D_{AB_n} *vs.* X. This pH range is obtained by preparing the reactant solutions in 0.002 (M) HCl.

Procedure

1) Prepare 10 ml mixtures of Fe (III) and salicylic acid solutions in labeled stopper glass tubes using burettes in the following ratios (Table **7.10**).

Table 7.10. Composition of Solutions.

Ferric Alum Solution (ml)	Salicylic Acid Solution (ml)
9	1
8	2
7	3
6	4
5	5
4	6
3	7
2	8
1	9

2) Scan the solution in the visible region taking water as blank, to find the value of λ_{max}. Measure the absorbance of each mixture at the same wavelength.

3) Plot the absorbance or transmittance of each mixture (ordinate) *vs.* x (zero to 1) and find the value of x corresponding to the maxima or minima of the curve. Calculate the value of n and find the formula of complex (in this particular case the value of n = 1, so the formula of the complex is: $Fe[C_6H_4(OH)COOH]^{3+}$.

Calculations

Prepare a series of solutions of ferric alum in 0.002 (M) HCl to find the stability constant of the complex. Saturate each solution with salicylic acid powder to convert the ferric ions completely into a complex form. The concentration of the complexion in each solution will be equal to that of ferric ions. Find the absorbance of each of these solutions and plot the calibration curve for the complex (by plotting absorbance or transmittance as ordinate against concentration).

Table 7.11. Table for stability constant.

S. No.	Initial Concentration of Fe (III) (a mol/lit)	Initial Concentration of Salicylic acid (b mol/lit)	Equilibrium Concentration of Complex (c mol/lit)	Equilibrium Concentration of Fe (III) ((a-c) mol/lit)	Equilibrium Concentration of Salicylic acid ((b-c) mol/lit)	Stability Constant (K_C)
1						
2						
3						
4						
5						
6						Mean=

The equilibrium concentration of the complex ions in each solution can be obtained corresponding to the observed absorbance from the calibration curve. Equilibrium concentrations of Fe (III) and salicylic acid in each solution can be calculated by subtracting the concentration of the complex from their respective initial concentrations (Table **7.11**).

IV. TO DETERMINE THE DISSOCIATION CONSTANT OF A GIVEN INDICATOR

Chemicals and Apparatus Required

0.2 (M) NaOH (Free from carbonate), 0.2 (M) Free broic acid, saturated solution of sodium carbonate, buffer solutions in pH range 8.0-10.4 made from NaOH and boric acid, freshly prepared solution of phenolphthalein (0.2 g in 100 ml rectified spirit, 10 ml of this solution is diluted to 100 ml with water), spectrophotometer, test tubes, pipette.

Theory

Acid-base indicators have different colours in acetic and alkaline solutions. Some examples are shown in Table **7.12**.

Table 7.12. Some common indicators and their colour in acidic and basic medium.

Indicator	pH at which Colour Change	pK_{In}
Phenolphthalein	8.2	9.7
Methyl red	red in pH under 4.4, yellow in pH over 6.2, and orange in between	5.1
Methyl orange	4.3.	3.7

The exchange of H^+ ions between the two forms causes colour change. Suppose the acid form of the indicator is represented by HIn, then its dissociation can be taken as

$$HIn \leftrightharpoons H^+ + In^-$$

The dissociation constant of the indicator is given by K_{In}, which is expressed as:

$$K_{In} = \frac{c_{H^+} c_{In^-}}{c_{HIn}} \quad \textbf{(7.38)}$$

If we express the concentration of the indicator in a solution as C moles/lit and its degree of dissociation as α, then

$$c_{H^+} = C\alpha = C_{In^-} \quad \text{and } c_{HIn} = C(1-\alpha) \quad \textbf{(7.39)}$$

The degree of dissociation can be shifted by changing the concentration of the indicator as well as by changing the pH. The pH can be changed by using appropriate buffer solutions. The changes in the colour of the indicator are perceptible from 10% dissociation to 90% dissociation of the indicator. This complete range is controlled by a change in pH of the solution in the range $pK_{In} \pm 1$. For finding the dissociation of an indicator experimentally, the buffers of pH range $pK_{In} \pm 1$ are required. When H ion concentration is controlled then the value of dissociation constant of the indicator can be expressed as:

$$K_{In} = \frac{C\alpha . C_{H^+}}{C\,(1-\alpha)} = \frac{\alpha . C_{H^+}}{(1-\alpha)} \quad \textbf{(7.40)}$$

Taking the reciprocals, we have:

$$\frac{1}{K_{In}} = \frac{1}{C_{H+}}\frac{1-\alpha}{\alpha}$$

Taking the log of base 10, we have:

$$pK_{In}=pH-\log\left(\frac{\alpha}{1-\alpha}\right)$$
$$\text{or, } \log\left(\frac{\alpha}{1-\alpha}\right)=pH-pK_{In}$$

$$pK_{In} = pH - \log\left(\frac{\alpha}{1-\alpha}\right)$$

$$\text{or, } \log\left(\frac{\alpha}{1-\alpha}\right) = pH - pK_{In} \quad \textbf{(7.41)}$$

The term pK_{In} is constant. The value of α increases with increasing the pH of the solution. For evaluation of dissociation constant of indicator, a known adjustment in the concentration of H^+ ion is involved as well as a measurement of the concentration of indicator ion or undissociated indicator or both of which may be coloured. The estimation of the concentration of dissociated and undissociated forms of the indicator can be done spectrophotometrically.

Procedure

1) Prepare buffer solutions as listed in Table **7.13**.

Table 7.13. List of buffer solutions prepared in test tubes.

Solution no.	**1**	**2**	**3**	**4**	**5**	**6**
0.2 (M) NaOH(ml)	0	1.0	2.0	3.0	3.5	4.0
0.2 (M) H_3BO_3(ml)	10.0	9.0	8.0	7.0	6.5	6.0
pH	6.90	7.95	8.54	8.98	9.12	9.45

2) 10 ml of saturated Na_2CO_3 is taken in the 7th test tube, which has a high pH value and causes almost complete dissociation of phenolphthalein.

3) Add 3 drops of indicator to each of the 7 test tubes. Maximum colour will be developed in test tube no. 7. This solution is used to find the value of λ_{max}. The

value of absorbance and transmittance is recorded for all the seven solutions at the λ_{max}.

Calculation

If C is the total concentration of phenolphthalein in each tube, the concentration of ionized phenolphthalein will be αC. The value of $\alpha = 1$ for the solution in the 7th test tube

$$\alpha_i C \propto A_i \text{ (For each test tube)}$$

$$\alpha_i C \propto A_7 \text{ (For test tube no. 7)}$$

A_i is the absorbance of the solution in the ith test tube.

For any test tube, $\alpha_i = \frac{A_i}{A_7}$

The absorbance of ith solution can be expressed as $A_i = \log\frac{I_0}{I_i} = \varepsilon\alpha_i Cx$.

$$A_i = \log\left(\frac{I_0}{I_i}\right) = \varepsilon\alpha_i CX$$

Here, I_0 and I_i are the intensity of incident radiation and intensity of radiation after passing through the ith solution.

For the 7th test tube $A_7 = \log\frac{I_0}{I_7} = \varepsilon Cx$

$$A_7 = \log\left(\frac{I_0}{I_7}\right) = \varepsilon CX$$

$$\alpha_i = \frac{\log\left(\frac{I_0}{I_i}\right)}{\log\left(\frac{I_0}{I_7}\right)} = \frac{\log\left(\frac{1}{T_i}\right)}{\log\left(\frac{1}{T_7}\right)}$$

$$\text{or,} \alpha_i = \frac{-\log T_i}{-\log T_7}$$

$$\alpha_i = \frac{\log\left(\frac{I_0}{I_i}\right)}{\log\left(\frac{I_0}{I_7}\right)} = \frac{\log\left(\frac{1}{T_i}\right)}{\log\left(\frac{1}{T_7}\right)} = \frac{\log T_i}{\log T_7} \qquad \textbf{(7.42)}$$

This shows how α can be calculated for each tube. Thus, pK_{In} can be calculated from Equation **(7.41).**

A calibration curve for phenolphthalein ions can be plotted by taking 10 ml of saturated Na_2CO_3 solution in 5 test tubes and adding 1, 2, 3, 4, and 5 drops of phenolphthalein solution, respectively. The value of absorbance for each solution is measured and plotted against the concentration of phenolphthalein ion. The calibration curve thus obtained can be used for determining ionized phenolphthalein in the test tubes taken in the first experiments. These observations can be used for calculating the value of the degree of dissociation (α) and pH and K_{In}.

FURTHER READING

[1] Source: http://ucdavis.edu
[2] Palit S. R.; De, S. K. "*Practical Physical Chemistry*" Science Book Agency, Calcutta, **1974**.
[3] Daniels, von F.; Mathews, J. H.; Williams, J.W.; Bender, P.; Alberty, R. A.; "*Experimental Physical Chemistry*" McGraw-Hill Book Company INC., New York-Toronto-London, **1956**.
[4] Athawale, V. D.; Mathur, P. *"Experimental Physical Chemistry" New Age International (P) Limited Publishers, New Delhi,* **2011**.

CHAPTER 8

Measurements Based on Electrode

pH-METRY

Introduction

Determination of pH is a very common measurement in laboratories, industry, hospitals, soil and water testing, and a wide variety of other analytical, control, and monitoring operations. Therefore, a large variety of instruments are available commercially to measure, control, and monitoring pH operations. Measurement of pH gives a quantitative idea of the degree of acidity (or basicity)- concentration of H^+ - in a stationary or flowing sample fluid. This is often very crucial in decision making in quality control, process industries, food, beverages and fermentation industry, biological and microbiological operations, *etc.*

Sensing of pH is accomplished by a pH-sensitive glass in contact with the internal fill, a 7-pH buffer, and the external sample. The pH-sensitive glass develops potential as per Equations (8.1) and (8.3), by proton exchange between H_3O^+ in the aqueous solution and the hydrated gel layer of the glass. The protons enter the sites vacated by Na^+ and recombine with H_3O^+ in the hydrated gel layer.

The potential developed is proportional to the difference in logarithms of activity of H_3O^+ in solution and gel layer on both sides of the glass membrane. The log of hydrogen ion activity can be converted to pH using the definition of pH, to give Equation (8.6):

$$E_1 = K_g^1 + 0.1984(T + 273.16)\log\frac{a_1}{a_g^1}$$

$$E_1 = K_g^1 + 0.1984\,(T + 273.16)\log\left(\frac{a_1}{a_g}\right) \quad \textbf{(8.1)}$$

$$E_1 = K_g^1 + 0.1984(T + 273.16)\,[\log a_1 - \log a_g^1]$$

$$E_1 = K_g^1 + 0.1984\,(T + 273.16)\log(a_1 - a_g) \quad \textbf{(8.2)}$$

Charu Arora and Sumantra Bhattacharya

$$E_2 = K_g^2 + 0.1984(T + 273.16)\log\frac{a_2}{a_g^2}$$

$$E_1 = K_g^2 + 0.1984\,(T + 273.16)\log\left(\frac{a_2}{a_g^2}\right) \quad \textbf{(8.3)}$$

$$E_2 = K_g^2 + 0.1984(T + 273.16)[\log a_2 - \log a_g^2]$$

$$E_1 = K_g^2 + 0.1984\,(T + 273.16)\log(a_2 - a_g^2) \quad \textbf{(8.4)}$$

$$\text{If } K_g^1 = K_g^2;\, a_g^1 = a_g^2 K_g^1 = K_g^2;\, a_g^1 = a_g^2$$

$$E_2 - E_1 = 0.1984(T + 273.16)[\log a_1 - \log a_2]$$

$$E_2 - E_1 = 0.1984\,(T + 273.16)[\log a_1 - \log a_2] \quad \textbf{(8.5a)}$$

$$= 0.1984(T + 273.16)(pH_2 - pH_1)$$

$$= 0.1984\,(T + 273.16)(pH_2 - pH_1) \quad \textbf{(8.5b)}$$

$$= 0.1984(T + 273.16)(7 - pH_1)$$

$$= 0.1984\,(T + 273.16)(7 - pH_1) \quad \textbf{(8.6)}$$

The above series of equations also indicate that:

1. mV output of electrode decreases as the pH increases

2. mV output is zero at pH = 7.

3. mV output is (+)-ve below pH 7 and (–)-ve above pH 7.

4. Effect of temperature on mV output approaches zero as pH approaches 7.

5. At 25^0 C the output changes by 59.16 mV per pH unit (0.198 x 298.16)

The pH values of various foodstuffs have been represented in Fig. (**8.1**). The pH and corresponding acidic, basic, and neutral substances along with H^+ and OH^- ion concentration are given in Table **8.1**.

In laboratories, pH is measured with the help of a pH meter consisting of an electrode and microprocessor. Detail of various types of measurement electrodes is given in Table **8.2**. The Source of errors in measurement and their symptoms are summarised in Table **8.3**.

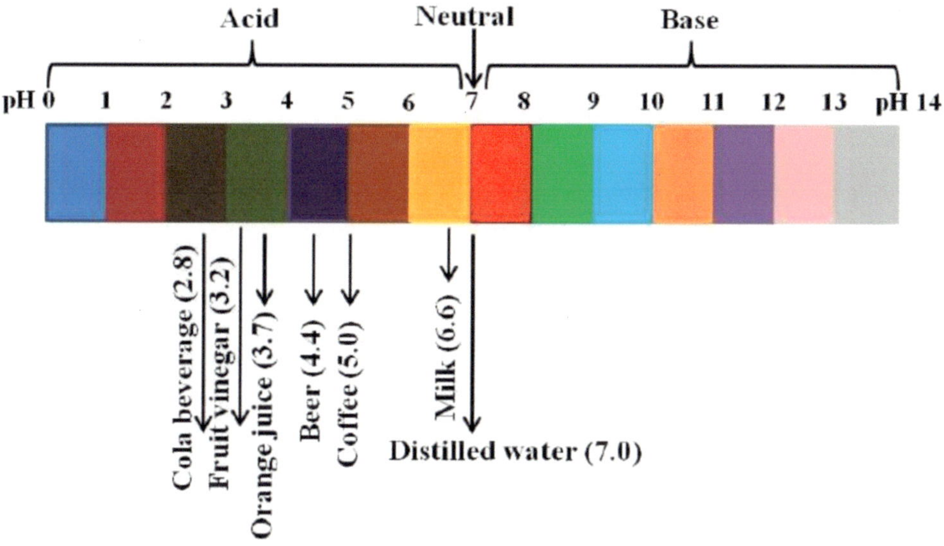

Fig (8.1). pH range of various food products.

Table 8.1. pH values of acidic, neutral and basic stuff.

Range	**pH**	**H^+ Concentration (mol/L)**	**OH Concentration (mol/L)**
Acid	0	1	0.00000000000001
	1	0.1	0.0000000000001
	2	0.01	0.000000000001
	3	0.001	0.00000000001
	4	0.0001	0.0000000001
	5	0.00001	0.000000001

	6	0.000001	0.00000001
Neutral	7	0.0000001	0.0000001
Alkaline	8	0.00000001	0.000001
	9	0.000000001	0.00001
	10	0.0000000001	0.0001
	11	0.00000000001	0.001
	12	0.000000000001	0.01
	13	0.0000000000001	0.1
	14	0.00000000000001	1

Table 8.2. The ratings of different types of measurement electrodes for various operating conditions. The + denotes a positive or good rating; the – denotes a negative or poor rating.

Type / Range	Accuracy	High Temp.	ORP	Sodium	Coating	Abrasion	Etching	Cost
Glass Bulb (0 to 14pH)	+	-	+	+	-	-	-	+
Flat Glass (2 to 12pH)	+	-	+	-	+	+	-	+
Glasteel	+	+	+	+	+	+	+	-

(0t o 10pH)								
Antimony (3 to 11pH)	-	-	-	-	-	+	+	+
Iridium Oxide (-10 to 20pH)	+	+	-	+	+	+	+	+
DURAFE T (0 to 14pH)	+	+	+	+	+	+	+	+

Table 8.3. Error and symptoms in pH Measurements.

Sources of Error	Electrical Symptom	Response Symptom	Effect on pH *vs.* mV Line
	Measurement Electrode		
Bulb Broken	$E_1 = E_2$ $R_1\downarrow$	No response	Horizontal line
Fill Contamination	$E_1 = E_2$ $R_1\downarrow$	-- do --	-- do --
Bulb Abrasion	$\Delta E_1/\Delta pH\downarrow$ $E_1\downarrow$	Slow, erratic, shorter span, unscale pH	Slope ↓, isopotential point shifts left
Bulb Dehydration	$E\downarrow$, $R_1\uparrow$ $\Delta E_1/\Delta pH\downarrow$	-- do --	-- do --
Bulb Etching	$\Delta E_1/\Delta pH\downarrow$, $E_1\downarrow$	-- do --	-- do --
Partial Bulb Coating	$\Delta E_1/\Delta pH\downarrow$	Very slow	Slope ↓
Complete Coating	E_1 fixed	No response	Horizontal line
Low temperature	$R_1\uparrow$	pH increases as temperature decreases	Isopotential point shifted ↓
	Reference Electrode		
Bulb Broken	$E_4\downarrow$ or ↑	Difference ↑ or ↓	Isopotential point down or up
	TERMINALS		
Short from M to R	$E_i = 0$	Fixed at 7 pH	Horizontal
Electrical Wire Broken	$E_i = 0$	-- do --	-- do --
Short from M to Gr	$E_i = 0$, $R_3 = 0$	-- do --	-- do --
Moisture on M	$R_3\downarrow$	Starys near 7 pH	**Slope ↓**
Moisture on R	$R_4\downarrow$	Upscale pH	Isopotential point shifts ↓

Glass Electrodes

A gel layer develops on the pH-sensitive glass membrane when a pH glass electrode is dipped into an aqueous measuring solution. Such a gel layer arises also on the inside the glass membrane which is in contact with a defined buffer solution (the inner buffer).

The protons (H^+ ions) either diffuse out of the gel layer or into the gel layer, depending on the pH value of the measured solution. In the case of an alkaline solution the ^-OH ions diffuse out and a negative charge is established on the outer side of the gel layer. Since the glass electrode has an internal buffer having a constant pH value, the potential at the inner surface of the membrane is also maintained during the measurement. The total membrane potential is an outcome of the difference between the inner and outer charges.

$$E_{el} = E^0 - S\,(pH_0 - pH_i) \qquad \textbf{(8.7)}$$

Here, E_{el}, E^0 is electrode potential and zero potential, respectively. S is the slope, expressed in the unit of mV/pH, pH_0 and pH_i are the pH value of the internal buffer and the same of the measured solution, respectively.

Combination Electrodes

Since the combination electrode is much easier to handle than the separate electrodes. In the combination electrode, the glass electrode is concentrically surrounded by the reference electrolyte. Only when the different parts of the electrode are expected to have very different life expectancies is the use of separate electrodes endorsed instead of a single combination electrode.

Three-in-one Electrodes

A recent innovation is the addition of a temperature sensor to the pH combination electrode. By casing the temperature sensor in the same body as the pH and reference elements, temperature compensated readings can easily be made with a single probe.

pH Measuring System

Successful pH measurement can only be accomplished by selecting the correct system to meet the demands of the sample under examination. As well as the correct apparatus, a supply of suitable reagents is vital.

Consideration has to be given to various parts and their utility in pH measurement (Table **8.4**).

Table 8.4. Various parts and their utility in pH measurement.

Type of pH Meter	Specification, Ease of Operation
Electrode(s)	Is it suitable for this measurement? Is a pH electrode with a built-in temperature sensor available?
Temperature probe	Is temperature compensation required?
Buffer solutions	The pure, correct value
Reagents	Distilled water, electrolyte solutions, cleaning solutions
Glassware	Clean, labeled
Electrode holder	For housing electrode(s)

pH Electrode System

Troubleshooting Guide

The first vital step in troubleshooting is to isolate the problem to one of the six major elements of the pH measuring system, which are:

1) pH meter

2) pH electrode

3) Reference electrode

4) Buffer solution

5) Operator

6) Application

Once this is done, action can be taken to correct the problem. Below we briefly discuss each of them

pH Meter

Symptoms usually related to pH failure are off-scale reading and reading that will not change. To confirm that the meter is the source of the problem, refer to the instruction manual accompanying the meter.

Electrodes

There are three types of electrodes used to measure pH:

(i) Reference Electrodes

(ii) pH Electrodes

(ii) Combination Electrodes (pH and reference electrodes in one sensor)

i. Reference Electrode

Symptoms usually related to reference electrode failure are inaccuracy or sluggish response, noisy or unstable readings, and off-scale readings.

Carry out the following check procedure:

1. Check that the filling solution is above the internal elements (not applicable for gel-filled electrodes).

2. Check side aperture is open (if applicable).

3. Look for signs of blockage or discoloration of reference junction (if the electrode has a replaceable ceramic junction this can be replaced as instructed).

4. Connect a working pH half-cell and the reference electrode to the pH meter. In mV mode, the system should display a stable reading. Drift or noise indicates an unshielded cable or a poor connection. If the reference electrode does not meet the above checks it should be cleaned thoroughly.

Proceed as Follows

1. Empty the reference chamber, rinse with deionized water, empty, and refill with the specified filling solution.

2. Soak the electrode in hot (50°C - 60°C) reference electrolyte for a few minutes or until the filling solution flows freely.

3. Soak electrode junction overnight in pH 4 buffer.

4. Remove any exterior salt deposits with distilled water.

5. If the filling solution does not flow through the junction by this time (generally due to unusually low junction porosity) the following procedure should be followed:

i) Hang the electrode in the air for some time and let KCl creep out and crystallize.

ii) Use gentle suction to pull the filling solution through if necessary.

iii) Repeat from step 2.

iv) Try the check procedure again.

6. Sometimes the material clogging the junction requires more severe action. Should the above failure, proceed as follows:

i) Use a solvent specific to the solution or material plugging the junction, if possible.

ii) Soak membrane overnight in 0.1 M HCl.

iii) If measurements were made in solutions containing protein or sulfides, remove deposits by soaking electrodes in an appropriate deproteinizing cleaning solution.

iv) Repeat from step 1.

7. Soak the electrode in an ammonium bifluoride regeneration solution.

If all these fail, the electrode should be replaced with a new electrode.

ii. pH Electrode

Symptoms usually related to pH electrode failure are noisy or unstable reading, off-scale readings, and 2-point calibration that cannot be performed.

To ascertain whether the pH electrode is at fault the following procedure should be followed:

1. With the meter set to read absolute mV, dip the pH electrode in question, with a working reference, in pH 7.00 buffer.

2. The readings should be 0 mV ± 30 mV with an Ag/AgCl reference. (This is checking the zero potential).

3. Lower the electrodes into pH 4.00 or pH 10.00 buffer and the reading should be greater than 150 mV different from the zero potential.

If the electrode responds outside of this test, clean the electrode as follows:

1. Degrease the membrane with cotton wool soaked in acetone (**CAUTION –**

HAZARDOUS SUBSTANCE) or soap solution.

NOTE: Acetone should not be used on plastic-bodied electrodes.

2. Soak membrane in 0.1 M HCl overnight.

3. If measurements have been made in samples containing protein, remove protein deposits by soaking electrode bulb in 0.1 M HCI + pepsin solution.

4. Soak the electrode in an ammonium bifluoride regeneration solution.

If all these fail, the electrode should be replaced with a new electrode.

iii. Combination Electrode

If a faulty combination electrode is suspected, a mixture of steps described for pH and reference electrodes should be tried. Commonly the reference is at fault and the reference procedure should be tried first. If this fails to return the electrode to the expected level of performance replace it with a new electrode. For short-term storage, the electrode should be immersed in a 15 mm depth of reference filling solution. Ensure that the level of solution in the beaker is below that of the filling solution in the electrode.

For longer-term storage, the wetting cap, filled with solution, should be replaced and the side filling aperture closed.

Other Sources of Error

A. Buffers

Symptoms usually related to buffers are inaccurate readings or the inability to perform 2-point calibration. Check for aged or contaminated buffers and rectify by substituting fresh ones.

B. Operator

Symptoms usually related to operator error are off-scale reading, noise, inability to perform 2-point calibration, and inaccuracies of reading. Check for the following common operator faults: Ensure that the unit is properly grounded or plugged into the wall outlet.

Ensure that the electrodes are plugged into proper terminals and are sealed firmly.

Ensure that the calibration of the meter is being performed according to the Operating instructions outlined in the Instruction Manual.

Ensure that the meter is being calibrated to the actual buffer required and that the buffers have not been reversed.

Check that the reference electrode has not been filled with the wrong filling solutions. (Calomel electrode solution becomes very milky if AgCl solution has been used). Before making a measurement check that the wetting cap and side filling aperture have been removed and **rinse the electrodes before** measuring a different buffer or sample.

Note - A short time spent reading the electrode product insert and the instruction manual for the instrument being used should eliminate most operator errors.

C. Applications

Symptoms usually related to applications problems are drift, slow response, and an unusual number of electrode failures.

Examples of typical application type problems are:

1. Electrodes in unbuffered solutions such as distilled water responds more slowly, appearing to drift. In these solutions, stability may not be achieved for three or four minutes.

2. Some applications required a particular type of electrode to be used, and if the correct one is not used a large number of electrode failures will result – contact the manufacturer for advice on applications.

3. The use of Ag/AgCl reference electrodes in a sample that contains halides.

Electrode Storage

Directly offer use before a period of storage the electrodes should be rinsed thoroughly with distilled water.

I. TO FIND OUT THE pK$_a$ VALUE OF GIVEN OXALIC ACID BY TITRATING IT AGAINST NaOH USING pH METER AT ROOM TEMPERATURE

Chemicals and Apparatus

Oxalic acid, NaOH, buffer solutions of pH 4 and 7. pH meter with electrodes, beaker, volumetric flask, burettes, pipettes, soft-tissue paper.

Theory

Oxalic acid is neutralised in two steps:

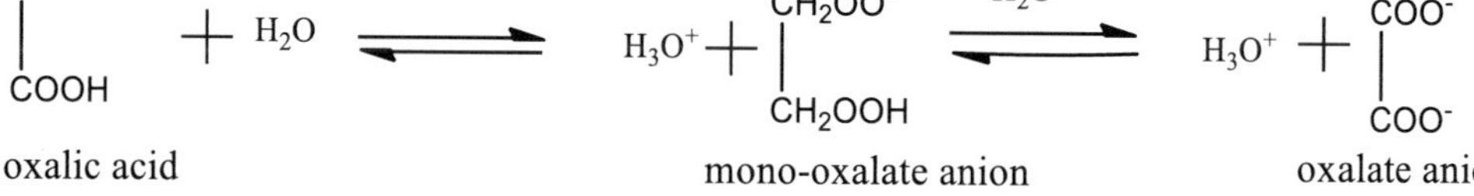

oxalic acid mono-oxalate anion oxalate anion

The first dissociation is approximately 10^3 times faster compared to the second step. Therefore, two neutralization points are expected on titrating it with a solution of NaOH. The pH of the solution at the half-neutralisation point of the first and half-neutralisation point of the second is designated as pK_{a_1} and pK_{a_2}, respectively. When the unknown volume of oxalic acid is titrated against NaOH pH-metrically and a plot of pH *vs.* volume of NaOH added is obtained is shown in Fig. (**8.2**).

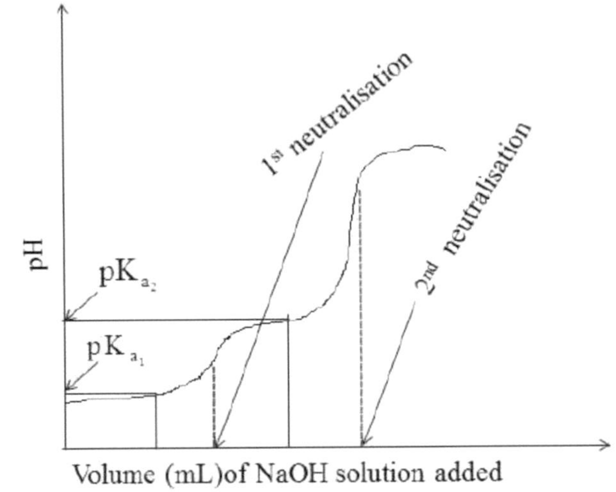

Fig. (8.2). Plot of pH *vs.* volume of NaOH solution added for titration of oxalic acid against NaOH solution.

The use of glass electrode and saturated calomel electrode (SCE) or combined pH electrode in the pH meter and its calibration is as usual. pK_a values can be determined from the pH values corresponding to half-neutralisation points of first and second neutralisation points, respectively (As indicated in the plot).

The strength of oxalic acid can be determined from the volume of NaOH solution required corresponding to the second neutralisation point.

The strength of NaOH can be determined from the volume of NaOH required corresponding to the second neutralisation point. To get an accurate neutralisation point, it is advisable to plot $\Delta pH/\Delta V$ *vs.* volume of NaOH solution added. It will give a two-sharp peak curve. The first small peak corresponds to 1st neutralisation point, while another peak indicates the second neutralisation point (Fig **8.3**).

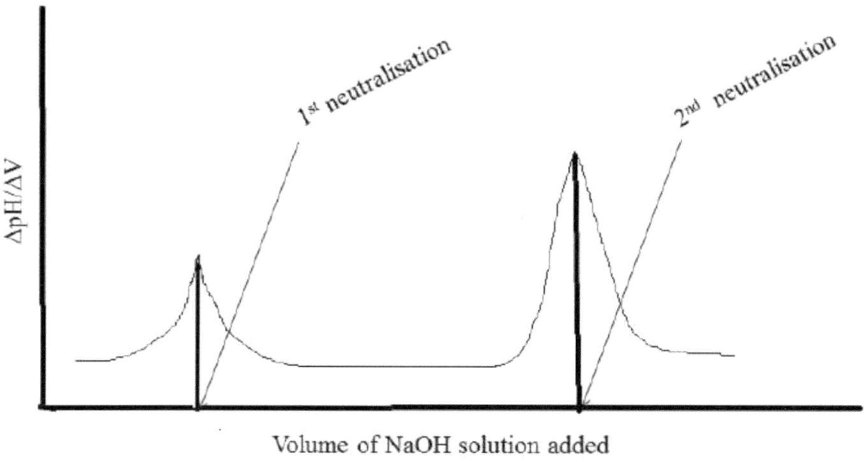

Fig. (8.3). Plot of $\Delta pH/\Delta V$ *vs.* volume of NaOH solution added.

Procedure

1) Prepare 100 ml 0.2 (M) oxalic acid solution by accurate weighing and ~100 ml 0.1 (M) NaOH solutions (Table **8.5**).

2) Calibrate the pH meter with buffer solutions of pH 4 and pH 7.

3) Take 25 ml of oxalic acid solution in a beaker and immerse the electrode into the solution. Observe initial reading of pH meter. Continue recording the pH value after the addition of 0.5 ml NaOH solution (from burette) after proper mixing (Table 8.6).

4) Take 5-6 more readings after complete neutralisation point (In the alkaline pH range, *i.e.*, pH>7.5).

5) Plot pH *vs.* volume of NaOH added to determine pK_a values as described in theory. Plot $\Delta pH/\Delta V$ *vs.* volume of NaOH solution added and determine the volume of NaOH required for complete neutralisation of 25 ml oxalic acid.

6) Calculate the strength of NaOH solution from the plots.

Observation and Results

Temperature =......° C

Table 8.5. Preparation of 100 ml 0.2 (M) oxalic acid solution.

Molecular weight of oxalic acid ($C_2O_4H_2.2H_2O$=126)

Initial Mass (g)	Final Mass (g)	Mass of Oxalic Acid transferred (g)	Mass of Oxalic Acid to be Taken (g)	Strength of Oxalic Acid Solution (M)
			2.52	

Table 8.6. Titration of oxalic acid solution pH-metrically.

Volume of oxalic acid solution taken = 25 ml.

S. No.	Volume of NaOH added (V mL)	Observed pH	ΔpH/ΔV
1	0		
2	1		
3	3		
4	4		
.	.		
.	.		
.	.		
.	.		

Calculation:

From the plot of pH *vs.* V

pK_{a_1}=……, pK_{a_2}=……At …...° C

From the plot of ΔpH/ΔV *vs.* ΔV

The volume of NaOH solution consumed = ... ml.

Strength of NaOH solution = (25×strength of oxalic acid solution)/(Volume of NaOH solution consumed).

II. TO FIND OUT THE STRENGTH OF HCL AND ACETIC ACID IN THEIR MIXTURE BY TITRATING IT AGAINST NaOH SOLUTION

Chemicals and Apparatus

HCl solution, acetic acid, pH meter, beaker, burette, pipette, measuring cylinder.

Theory

pH is defined as the negative logarithm of the activity of H^+ ions. Mathematically, $pH = -\log_{10} a_{H^+}$. For considerably dilute solution, activity is replaced by concentration. Thus, the expression of pH becomes: $pH = -\log c_{H^+}$.

The pH of a solution can be determined accurately using a glass electrode consisting of a thin membrane of soda-lime glass containing dilute HCl acid, in which the Ag-AgCl electrode is immersed. This electrode is coupled with the reference electrode, *viz.* SCE. The cell obtained is:

Ag-AgCl|0.1HCl|glass|unknownsolution|SCE

The EMF of the above cell can be expressed as:

$$E = E_G - E_{SCE} = E_G^0 - E_{SCE} + \frac{2.303RT}{F} pH$$

$$E = E_G - E_{SCE} = E_G^0 - E_{SCE} + \frac{2.303RT}{F} pH \quad \textbf{(8.8)}$$

The emf is a linear function of the pH of the solution in which the glass electrode is kept immersed. The resistance of the glass membrane is considerably high (>1 Ω), the EMF of the cell cannot be measured using an ordinary potentiometer. A pH meter is required for the measurement of emf of the cell. This experiment can be carried out into two parts.

1) Standardisation of NaOH solution using the standard oxalic acid solution: Same as above experiment.

2) Determination of strength of HCl and acetic acid: The acid mixture is titrated with standard NaOH solution. During titration, the pH of the solution increases by the addition of NaOH. Since HCl is stronger acid compared to the other one, so due to the common ion effect, HCl will be in ionised form, hence HCl will be neutralised first. After complete neutralisation of HCl, the acetic acid gets neutralised. The titration is carried out till the solution achieves pH 12. A plot of ΔpH/ΔV *vs.* V is plotted as shown in Fig. (**8.4**).

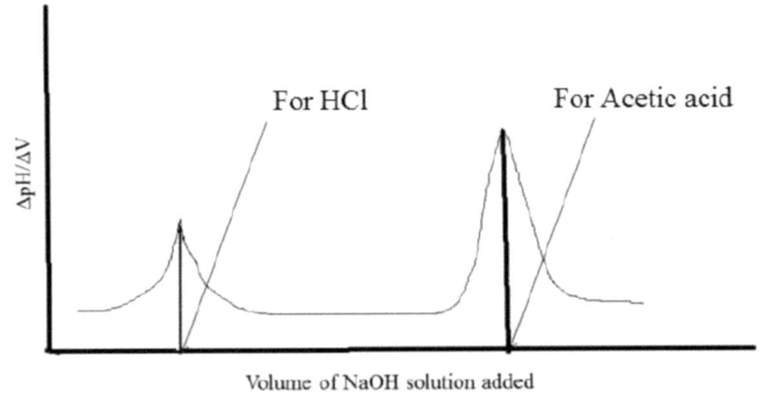

Fig. (8.4). Plot of pH metric titration of acid mixture by NaOH.

So, knowing the volume of standard NaOH consumed for neutralisation of HCl and acetic acid we can calculate individual strengths of HCl and acetic acid in their mixture. If V_1 is the volume of NaOH consumed for HCl and V_2 is the same for total acid mixture, then the volume of standard NaOH solution required for neutralisation of acetic acid will be (V_2- V_1).

Procedure

1) Prepare 100 ml 0.1 (M) oxalic acid solution by accurate weighing (Table **8.7**).

2) Prepare 100 ml ~0.5 (M) NaOH solution.

3) Standardisation of NaOH solution: The procedure is the same as experiment II. The strength of NaOH is determined by the graphical method as described in experiment II (Table **8.8**).

4) Titration of acid mixture: Wash the electrodes thoroughly and dip them in 10 ml of the acid mixture (supplied) in a 50 ml beaker and continue titration till pH of solution achieves 12 (Table **8.9**). Determine the strength of HCl and acetic acid in the mixture graphically.

5) Wash and clean the electrodes and immerse them into fresh distilled water or saturated KCl solution.

Observation and Results

Temperature = ...° C

Table 8.7. Preparation of 100 ml 0.1 (M) oxalic acid solution.

Molecular weight of oxalic acid ($C_2O_4H_2$. $2H_2O$=126)

Initial Mass (g)	Final Mass (g)	Mass of Oxalic Acid Transferred (g)	Mass of Oxalic acid to be Taken (g)	Strength of Oxalic Acid Solution (M)
			1.26	

Table 8.8. Standardisation of NaOH solution pH-metrically.

Volume of oxalic acid solution taken = 10 ml.

Sl. No.	Volume of NaOH added (V mL)	Observed pH	ΔpH/ΔV
1	0		
2	1		
3	3		
4	4		
.	.		
.	.		

.	.		
.	.		

From the plot, the strength of NaOH is = …. (M)

Table 8.9. Titration of acid mixture pH-metrically.

Volume of oxalic acid solution taken = 10 ml.

Sl. No.	Volume of NaOH added (V mL)	Observed pH	ΔpH/ΔV
1	0		
2	1		
3	3		
4	4		
.	.		
.	.		
.	.		

Determine V_1 (volume of NaOH required for neutralisation of HCl) and V_2 (volume of NaOH required for neutralisation of acid mixture). The volume of NaOH required for neutralisation of acetic acid= (V_2- V_1) ml.

Strength of HCl = …... (M) = …. g/l

Strength of Acetic acid = …... (M) = …. g/l

III. DETERMINATION OF THE DISSOCIATION CONSTANT OF ACETIC ACID USING pH-METER

Chemicals and Equipment

Acetic acid, sodium hydroxide, pH meter, 100 cm^3 beaker, glass electrode, calomel electrode.

Theory

The strength of an acid is experimentally measured by determining its electrolytes; they are ionized almost completely in aqueous solutions. It is not meant to study the ionic equilibrium of strong acids and calculate their equilibrium constants as the unionized form is present to such a small extent. Hence, the study of ionic equilibrium and calculation of K applies only to weak acids.

e.g., Acetic acid ionizes feebly as:

$$CH_3COOH\ (aq) + H_2O\ (l) \rightarrow H_3O^+\ (aq) + CH_3COO^-\ (aq);$$

$$K = \frac{C_{CH_3COO^-} \times C_{H_3O^+}}{C_{CH_3COOH}} \quad \textbf{(8.9)}$$

pK_a is a modern method of expressing the acid strengths or dissociation constant of an acid. pK_a is determined by measuring the changes in pH of the acid solution at different amounts of the base added, first slowly, then more rapidly, till the equivalence point is achieved. The equivalence point can easily be identified by a very sharp increase in pH for a very small quantity of added base. Once past the equivalence point, the pH increases only slightly on the addition of excess base. The titration curve is obtained by plotting changes in pH at different amounts of the base added and the equivalence point is determined.

Procedure

Pipette out 50 cm^3 of the given weak acid into a 100 cm^3 beaker. Immerse a glass electrode calomel electrode assembly into the acid and connect the cell to a pH meter. Measure the pH of the acid. Fill a burette with the base (sodium hydroxide). In the beginning, add large increments of (say 1 cm^3) of the base to the acid. Stir the solution thoroughly and measure the pH after each addition. When the pH begins to show a tendency to increase rapidly, add only small increments (say 0.1 cm^3) of the base and measure the pH after each addition. Continue till there is only a slight increase in pH on the addition of the base. Plot a graph of pH (ordinate) against the volume of sodium hydroxide added (abscissa). Determine the equivalence point and hence the pH at the half equivalence point. This gives the pKa value of the acid.

Observations and Calculations

Temperature = ----° C

Equivalence point = --------

Half-equivalence point = ------

pH at half-equivalence point = -----

pKa of the given weak acid = pH at the half equivalence point

Thus, pK_a of acetic acid at ° C = -------

CONDUCTOMETRY

I. TO FIND OUT CELL CONSTANT OF THE CONDUCTIVITY CELL

Chemicals and Apparatus

KCl solution, Conductometer, conductivity cell, beakers, burette, thermometer.

Theory

Resistance (R) of a conductor is directly proportional to the length (l) of the conductor and inversely proportional to the cross-section (a) of the conductor. Thus, mathematically we can write:

$$R = \rho \frac{l}{a} \tag{8.10}$$

Here, ρ is the specific conductance.

The inverse of R is known as conductance (C). Thus, $C = R^{-1}$. So,

$$C = \rho^{-1} \frac{a}{l} = \kappa \frac{a}{l} \tag{8.11}$$

Here κ is specific conductivity. Thus, the specific conductivity of a solution is the conductivity of 1 cc of the solution. The observed conductivity will be equal to the specific conductivity if the distance between the electrodes is 1 cm and the area of cross-section is 1 cm^2. Maintaining unit length and the unit cross-section is not practical for all experimental conditions. So observed conductivity must be multiplied by factor l/a to get the value of specific conductivity. The values of l and a are constants for a particular cell, so l/a (constant quantity) is known as '**cell constant (K_c)**'.

Thus,

$$\kappa = c\,\frac{a}{1} = c\,K_c \tag{8.12}$$

$$K_c = \frac{\kappa}{c} \tag{8.13}$$

The cell constant is determined by taking a solution of known specific conductivity in the cell and from the observed conductivity measured the value of the cell constant can be determined. Solutions of KCl of known concentration and known conductivity are used for the determination of cell constant. The values of specific conductivity of KCl solution having different concentrations have been given in Table **8.10**.

Table 8.10. Specific conductivity of KCl solution in mho-cm^{-1}(Kohlrausch Table).

Temperature (°C)	(M)	0.1 (M)	0.02 (M)	0.01 (M)
0	0.0654	0.00716	0.001522	0.000776
5	0.0740	0.00822	0.001752	0.000895
10	0.0832	0.0932	0.001995	0.001019
15	0.0925	0.01048	0.002243	0.001147

20	0.1020	0.01167	0.002501	0.001278
25	0.1120	0.01289	0.002768	0.001412
30	-----	0.01412	0.003036	0.001552

Procedure

1) Prepare exactly 0.1 (M) 100 ml KCl solution.

2) Pipette out 10 ml of 0.1 (M) and dilute it to 100 ml to prepare exactly 0.01 (M) KCl solution (Table **8.11**).

3) Switch on the conductometer and keep it on for at least 15 min before taking a reading.

4) Wash the conductivity cell with conductivity water. To ensure proper cleaning of the cell, insert the cell in ~25 ml conductivity water and observe the reading. Repeat the process 2-3 times with the same volume of water and be sure to get the same consecutive values for conductance.

5) Rinse the electrodes with 0.01 (M) KCl solution. Insert the cell in 25 ml KCl solution. Keep the cell completely immersed in the experimental solution. If required a larger definite volume can be taken. Record the temperature of the solution and observe its conductance.

6) Repeat Step 5) with 0.1 (M) KCl solution with the same volume as was taken in step 5)

7) Find the value of specific conductance at the experimental temperature from Kohlrausch Table (Table 8.10) for 0.01 (M) and 0.1 (M) KCl solution. Calculate the cell constant of conductivity cell for two solutions using Equation (8.13). Take the mean of the two values (Table **8.12**).

Observation and Results

Temperature = ------- ° C

Table 8.11. Preparation of 100 ml 0.1 (M) KCl solution.

Weight of Empty bottle (w_1) g	Weight of Empty Bottle + KCl (w_2) g	Weight of KCl Taken (w_2-w_1) g	Strength of KCl Solution (M)
			((w_2-w_1)/0.7456) ×0.1

From this solution, prepare exactly 0.1 (M) KCl solution by exact dilution

Table 8.12. Determination of cell constant.

S. No.	Strength of KCl Solution (M)	Observed Conductance (mho)	Specific Conductance (at Experimental Temperature) (mho-cm^{-1})	Cell Constant $K_c = \frac{\kappa}{c}$ (cm^{-1})	Mean K_c cm^{-1}
1	0.01 (M)				
2	0.1 (M)				

Result

The cell constant of given conductivity cell is = ------ cm^{-1} at ------- ° C.

II. TO FIND OUT THE STRENGTH OF STRONG BASE BY TITRATING IT AGAINST STANDARD WEAK ACID SOLUTION AND TO FIND OUT STRENGTH OF STRONG ACID BY TITRATING IT AGAINST STANDARD STRONG BASE

Chemicals and Apparatus

Oxalic acid, NaOH, HCl, conductometer set up, burette, beaker, and pipette.

Theory

Oxalic acid is a weak dibasic acid. It dissociates successively in two stages, such that the first dissociation constant is much greater than that of the second dissociation constant.

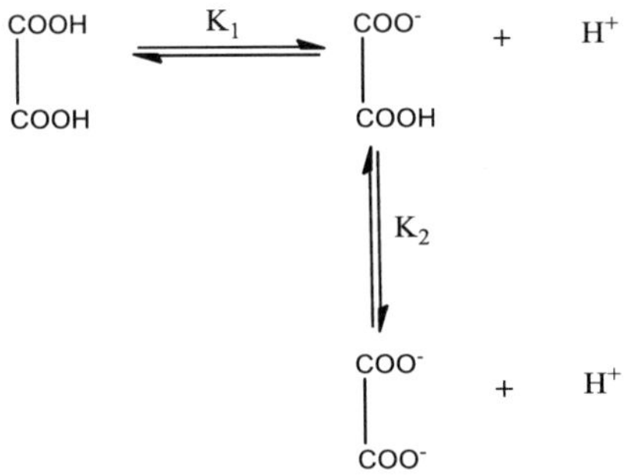

$K_1 \gg K_2$. Using conventional titrimetric methods, establishing two steps is too difficult. On the other hand, using instrumental methods, like pH-metry, potentiometry, conductometry, it can be easily established by the successive addition of NaOH. If we plot conductance *vs.* volume of NaOH added, two breaks will be obtained for complete titration. At the initial stage, conductance decreases rapidly (Sharp slope), after the first break (the break corresponds to 1st neutralisation) the curve rises slowly, due to the formation of mono acidic salt, which behaves like a weak base. After the end point (2nd break indicative of dibasicity), again conductance increases rapidly because of the increase of NaOH concentration in the solution.

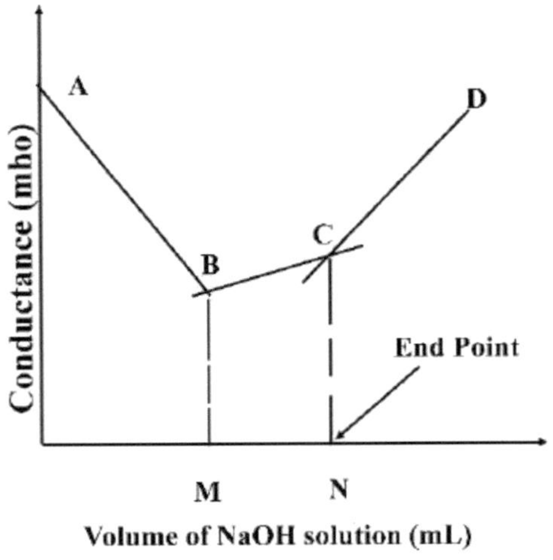

Fig. (8.5). Plot of weak dibasic acid *vs.* strong mono acidic base.

However, the plot for titration of HCl and NaOH is not the same. Here one should expect the following curve. The break point indicates the end point of the titration process.

From the first graph, we can get the strength of NaOH (2nd break) (Fig. **8.5**). Similarly, with the help of the breaking point of the second graph, HCl strength can also be calculated (Fig. **8.6**).

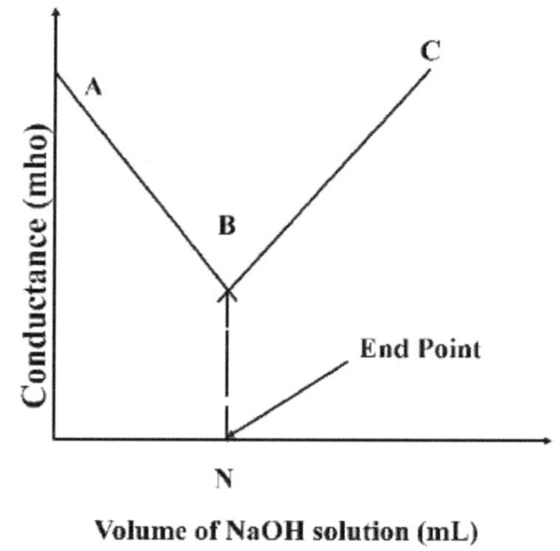

Fig (8.6). Plot of weak monobasic acid *vs.* strong mono acidic base.

The big advantages of conductometric titrations are:

1) It is more accurate compared to the conventional titrimetric method, because no indicator is used, so consumption of H^+ and OH^- by indicator is nil.

2) Dilution does not affect the end point for conductometric titration. Because dilution can never change the number of ions.

Procedure

1) Wash the conductivity cell thoroughly with conductivity water.

2) Prepare 50 ml 0.1 (M) standard oxalic acid solution by exact weighing and get the exact strength (Table **8.13**). Also, prepare ~ 0.1 (M) each of NaOH and HCl solution.

3) Pipette out 10 ml oxalic acid in a 100 ml beaker and insert the cell in it. Conductivity water can be added if required to immerse the electrodes completely in the solution.

4) Connect the cell with a conductometer and take readings of the conductance of oxalic acid solution.

5) Fill up the burette with NaOH solution. Start adding NaOH solution to oxalic acid in the beaker and shake gently for 30 sec and record conductance. Continue the addition process till we get a rapid increase in conductance (Table **8.14**).

6) Plot conductance *vs.* volume of NaOH added. Calculate the strength of NaOH with the help of a second break (end point).

7) Repeat steps 1) - 6), but take 10 ml of HCl, instead of 10 ml of oxalic acid and find the strength of HCl from the break point (Table **8.15**).

Observations and Results

Temperature = ...° C

Table 8.13. Preparation of 50 ml 0.1 (M) oxalic acid solution.

Molecular weight of oxalic acid ($C_2O_4H_2.2H_2O$=126)

Initial Mass (g)	Final Mass (g)	Mass of Oxalic Acid Transferred (g)	Mass of Oxalic Acid to be Taken (g)	Strength of Oxalic Acid Solution (M)
			0.63	

Table 8.14. Conductometric titration of standard oxalic acid solution *vs.* NaOH.

Amount of oxalic acid taken = 10 ml.

Serial no.	Volume of NaOH Added (ml)	Conductance (mho)
1	0.1	
2	0.2	

3	0.3	
4	0.4	
5	0.5	
6	0.6	
7	0.7	

NaOH consumed (V_1) = …… ml. Volume of oxalic acid (V_2) = 10 ml, strength of oxalic acid (S_2) = 0.1 (M). Thus, strength of NaOH (S_1) =10×0.1/S_2

Table 8.15. Conductometric titration of HCl solution of unknown strength *vs.* standardised NaOH solution.

Volume of HCl taken = 10 ml.

S. No.	Volume of NaOH Added (ml)	Conductance (mho)
1	0.1	
2	0.2	
3	0.3	
4	0.4	
5	0.5	
6	0.6	
7	0.7	

NaOH consumed (V_3) =…… ml. Volume of HCl acid (V_4) = 10 ml, strength of NaOH (S_3) = 10×0.1/S_2 (M). Thus, strength of HCl (S_4) = (10×0.1/S_2) × V_3/10 (M)

Thus, the strength of supplied HCl solution is = …... (M).

Precautions

Do not keep the electrode in the open air. Always dip in conductivity water or saturated KCl solution when it is not in use.

III. TO FIND OUT THE STRENGTH OF A MIXTURE OF ACID BY TITRATING IT AGAINST A STRONG BASE

Chemical and Apparatus

HCl and acetic acid, NaOH, oxalic acid conductometer set up, burette, pipette, and beaker.

Theory

The titration of a mixture of acids is like a combination of two separate titrations, *viz*, HCl against NaOH and acetic acid against NaOH. When a strong alkali (NaOH) is added to the mixture, the conductivity of the solution decreases sharply due to the replacement of the H^+ ion from the strong acid. The dissociation of acetic acid is almost inhibited due to the presence of a large number of H^+ ions resulting from the complete dissociation of HCl. Hence the conductance of the mixture continues to form till the complete neutralisation of HCl.

After consumption of all H^+ ions of HCl dissociation of acetic acid starts and the addition of NaOH solution slowly increases the conductance due to the formation of highly ionizing salt sodium acetate. After complete neutralisation of acetic acid, the conductance of the mixture increases rapidly on further addition of NaOH. The nature of the graph is displayed in Fig. (**8.7**). Where the first break indicates the neutralisation of HCl and the second is that for acetic acid.

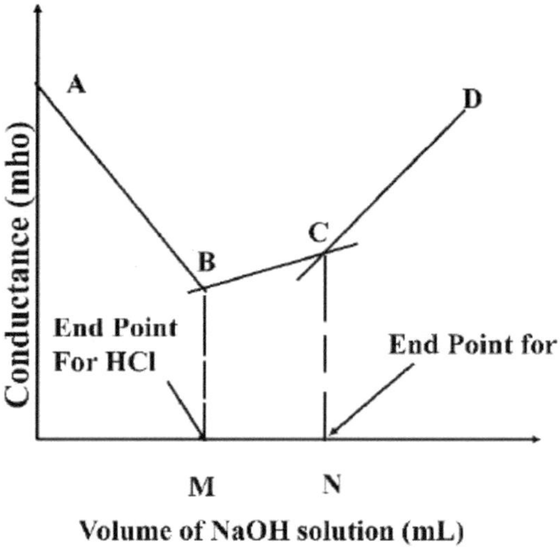

Fig. (8.7). Plot of titration of a mixture of acid against a strong base. OM => Volume of NaOH consumed for neutralisation of HCl, MN=> Volume of NaOH consumed to neutralise the acetic acid.

The total volume of NaOH solution required for complete neutralisation of the mixture-that required for HCl gives the volume of NaOH consumed for neutralisation of acetic acid.

Procedure

1) Prepare 50 ml 0.1 (M) standard oxalic acid solution (Table **8.16**).

2) Prepare 100 ml 0.1 (M) NaOH solution and standardise it against oxalic acid (Table **8.17**) conductometrically as described in Experiment II.

3) Take 10 ml of the acid mixture by pipette and standardise against standard NaOH solution adding small volumes of NaOH, record the conductance (Table **8.18**).

4) Plot the conductance *vs.* volume of NaOH added. Find the volume of NaOH required for neutralisation of HCl and acetic acid in the mixture.

Observation and Results

Temperature = ...° C

Table 8.16. Preparation of 100 ml 0.01 (M) oxalic acid solution.

Molecular weight of oxalic acid ($C_2O_4H_2.2H_2O$ = 126).

Initial Mass (g)	Final Mass (g)	Mass of Oxalic Acid Transferred (g)	Mass of Oxalic Acid to be Taken (g)	Strength of Oxalic Acid Solution (M)
			0.63	

Table 8.17. Conductometric titration of standard oxalic acid solution *vs.* NaOH (Amount of oxalic acid taken= 10 ml).

S. No.	Volume of NaOH Added (ml)	Conductance (mho)
1	0.1	
2	0.2	
3	0.3	
4	0.4	
5	0.5	
6	0.6	
7	0.7	
8	------	
9	------	
10	------	

NaOH consumed (V_1) =...... ml. Volume of oxalic acid (V_2) = 10 ml, strength of oxalic acid (S_2) = 0.1 (M). Thus, strength of NaOH (S_1) =10×0.1/S_2

Table 8.18. Conductometric titration of mixture of acids of unknown strength *vs.* standardised NaOH solution.

Volume of acid mixture taken = 10 ml.

S. No.	Volume of NaOH Added (ml)	Conductance (mho)
1	0.1	
2	0.2	
3	0.3	
4	0.4	
5	0.5	
6	0.6	
7	0.7	
8	------	
9	------	
10	------	

For Calculation of strength of HCl Solution (Consider up to 1st break)

NaOH consumed (V_3) = …… ml. Volume of HCl acid (V_4) = 10 ml, strength of NaOH (S_3) = $10\times0.1/S_2$ (M). Thus, strength of HCl (S_4) = $(10\times0.1/S_2) \times V_3/10$ (M)

For Calculation of strength of acetic acid solution (Consider from 1st break to 2nd break)

NaOH consumed (V_3') =…… ml. Volume of acetic acid (V_4) = 10 ml, strength of NaOH (S_3) = $10\times0.1/S_2$ (M). Thus, strength of acetic acid (S_4) = $(10\times0.1/S_2) \times V_3/10$ (M)

The strength of supplied HCl solution is = …... (M).

The strength of the supplied acetic acid solution is = …... (M).

IV. DETERMINATION OF SOLUBILITY AND SOLUBILITY PRODUCT OF A SPARINGLY SOLUBLE SALT

Chemicals and Apparatus

Sparingly soluble salt (either one of $BaSO_4$, AgCl, $PbSO_4$, potassium hydrogen tartrate, $PbCrO_4$), KCl, conductometer, reagent bottle, beaker.

Theory

Solubility of some salts like $BaSO_4$ or AgCl, $PbSO_4$, potassium hydrogen tartrate, $PbCrO_4$ are very less in water. They are known as sparingly soluble salts. Solubility is too small to be determined from ordinary analytical methods. However, it can be determined by conductometric measurements.

Let us assume a sparingly soluble salt B_xA_y dissociate as:

$$B_xA_y \rightleftarrows xB^{y+} + yA^{x-}. \quad \textbf{(8.14)}$$

The equilibrium constant:

$$K = \frac{a^y_{B^{x+}} a^x_{A^{y-}}}{a_{B_xA_y}} \quad \textbf{(8.15)}$$

For sparingly soluble salt, the denominator of the Equation (8.15) is 1. So, the equation becomes:

$$K_{SP} = a^y_{B^{x+}} a^x_{A^{y-}} \quad \textbf{(8.16)}$$

Here K_{SP} is the solubility product. Thus, the solubility product of a sparingly soluble salt is defined as the product of the activity of the constituent ions after raising them to appropriate power in the saturated solution of the salt at a particular temperature.

For low concentration, activity can be replaced by the concentration of the ions. Thus Equation (8.16) can be written as: $K_{SP} = c^y_{B^{x+}} c^x_{A^{y-}}$. If the solubility be x mol/lit for a sparingly soluble salt, then for 1-1 (like AgCl) or 2-2 ($BaSO_4$, $PbSO_4$) electrolytes $K_{SP} = x^2$. For 1-2 (Ag_2SO_4), 2-1 ($PbCl_2$) electrolytes, the value of solubility product is $K_{SP} = 4x^3$. Similarly, for other systems, the expression for K_{SP} changes accordingly.

The equivalent conductance (Λ) of an electrolyte is related to its specific conductance (κ) as:

$$\Lambda = \frac{1000\kappa}{c} \qquad \textbf{(8.17)}$$

Here c is the normality of the solution. Since the concentration of the saturated solution of a sparingly soluble salt is very low, equivalent conductance (Λ) may be replaced by the equivalence conductance at infinite dilution. Taking into account the specific conductance of water, the value of equivalent conductance for sparingly soluble salt, take the form:

$$\Lambda^0 = \frac{1000\,(\kappa_{salt} - \kappa_{water})}{c}$$

$$\text{Or, } c = \frac{1000\,(\kappa_{salt} - \kappa_{water})}{\Lambda^0}$$

$$\text{Or,} \quad c = \frac{1000\,(\kappa_{salt} - \kappa_{water})}{\Lambda_+^0 + \Lambda_-^0}$$

(8.18)

Λ_+^0 and Λ_-^0 are the ion conductance of the constituent ions of the salt. Hence, measuring the specific conductance of the saturated solution of a sparingly soluble salt (κ_{salt}) and putting the literature values of the specific conductance of water (κ_{water}), Λ_+^0 and Λ_-^0 in Equation (8.15), the normality (N) of the salt solution can be determined. If x mol/lit is the solubility of the salt, then for AgCl x=c, $BaSO_4$ x=c/2 or $Ca_3(PO_4)_2$ x=c/6, *etc.*

On putting the value of x in Equation (8.16), the K_{SP} of the salt can be calculated

Procedure

1) Record the experimental temperature.

2) Prepare 100 ml 0.1 (M) KCl solution to determine the cell constant of the conductivity cell.

3) Prepare a saturated solution of given sparingly soluble salt by shaking it for half an hour in conductivity water. Allow it to settle for 15 minutes and ensure that some salt appears as sedimentation. Filter the solution using Whatman 42 filter paper.

4) Record the conductance of the solution (c) and calculate the specific conductance as:

$$\kappa = c \times \text{cell constant}$$

5) Search the values of specific conductance of conductance water, the conductivity of Ag^+ and Cl^- ion at infinite dilution from the literature and fit them in Equation (8.16) to calculate the normality (N) of the saturated solution. Convert the normality into molarity which is the solubility of salt in terms of mole/lit.

6) Calculate K_{SP} from Equation (8.16).

Observation and Results

Temperature = ...° C

Please follow Tables **2** and **3** in Experiment (i) to determine cell constant and designate them here as Table **1** and **2**.

Table 8.19. Determination of solubility.

Conductance of the Solution (mho)	Cell const. (K_c) (cm^{-1})	Specific conductance (κ_{salt}) κ = conductivity × K_c ($mho\text{-}cm^{-1}$)	Specific Conductance of Water (κ_{water}) ($mho\text{-}cm^{-1}$)	$\Lambda^0_{salt} = \Lambda^0_+ + \Lambda^0_-$	(N)	Solubility (x) mole/lit

Result

From Table **8.19**, we can get the K_{SP} from Equation 8.16.

Thus, the Solubility of the given salt (******) is =.... at ....° C.

The solubility product of the given salt (******) is =...... at ....° C.

V. TO DETERMINE DISSOCIATION CONSTANT OF WEAK ACID AND VERIFY OSTWALD DILUTION LAW

Chemicals and Equipment

Glacial acetic acid, conductivity water, NaOH, oxalic acid, conductometer set up, beakers, volumetric flask, burette.

Theory

The ionisation of a weak electrolyte is represented as:

$$AB \leftrightharpoons A^+ + B^-$$

On applying the law of mass action, the dissociation constant (K) of the above equilibrium:

$$K = \frac{c_{A^+} c_{B^-}}{c_{AB}} \tag{8.19}$$

If α is the degree of dissociation of AB and we start with 1 mol of AB, then the amounts of A^+, B^- and undissociated AB at equilibrium will be α, α and (1-α), respectively. Let V L be the total volume of the system, then $C_{A^+} = \frac{\alpha}{V} = C_{B^-}$, $c_{AB} = \frac{1-\alpha}{V}$. Putting the concentrations of all the terms in Equation (8.19), we get:

$$K = \frac{\alpha^2}{(1-\alpha)V} \tag{8.20}$$

For a weak electrolyte, the value of the degree of dissociation is negligible in comparison to unity. Hence, $1-\alpha \approx 1$. Putting this value in Equation (8.20), we get:

$$K = \frac{\alpha^2}{V} \tag{8.21}$$

Equation (8.19) and Equation (8.20) are expressions for Ostwald's dilution law. If the degree of ionisation at a particular dilution is known, then, the value of K can be calculated from Equation (8.20) or (8.21). We know:

$$\alpha = \frac{\lambda_v}{\lambda_\infty} \tag{8.22}$$

Here λ_V and λ_∞ are equivalent conductance at dilution V and infinite dilution, respectively. From Kohlrausch's law, $\lambda_\infty = \lambda_a + \lambda_c$. Here, λ_a and λ_c are ionic conductance of the anion and cation of the substance being studied. These values can be obtained from the literature. Thus, putting the value we can get α as:

$$\alpha = \frac{\lambda_v}{\lambda_a + \lambda_c} \quad \textbf{(8.23)}$$

We find the dissociation constant of the given substance at various dilutions to verify Ostwald's dilution law. If the values come to be constant, then the law is verified.

Procedure

1) Determine the cell constant as done in Experiment I.

2) Prepare 100 ml 0.5 (M) solution of oxalic acid solution (Table **8.20**) by accurate weighing (Use conductivity water for solution preparation).

3) Prepare approximately 0.5 (M) solution of acetic acid from glacial acetic acid (Use conductivity water for solution preparation).

4) Prepare 100 ml 0.5 (M) NaOH solution and standardise it with oxalic acid solution conductometrically (Experiment (ii)) (Use conductivity water for solution preparation).

5) Following Experiment II also get the accurate strength of acetic acid.

6) Prepare exact 0.5 (M) acetic acid solution. Clean and dry the conductivity cell. Prepare (M/4), (M/8), (M/16), (M/32), (M/64), and (M/128) solutions of acetic acid from stock (M/2) solution by exact dilution with conductivity water.

7) Conductivity is measured for all the above acetic acid solution (Table **8.21**). Never forget to wash the cell thoroughly after each measurement.

Observations

Temperature = ...° C.

Cell constant = ...

λ_∞ for acetic acid = $\lambda_{H^+} + \lambda_{CH_3COO^-}$ = ...

Table 8.20. Preparation of 100 ml 0.5 (M) oxalic acid solution.

Molecular weight of oxalic acid ($C_2O_4H_2$. $2H_2O$ = 126)

Initial Mass (g)	Final Mass (g)	Mass of Oxalic Acid Transferred (g)	Mass of Oxalic Acid to be Taken (g)	Strength of Oxalic Acid Solution (M)
			6.3	

Table 8.21. Conductance Table.

S No.	Molarity of Acetic Acid	Observed Conductivity (Siemens per meter)	Equivalent Conductivity (λV) (Siemens per meter)	Degree of Dissociation (α)
1	(M/2)			
2	(M/4)			
3	(M/8)			
4	(M/16)			
5	(M/32)			
6	(M/64)			
7	(M/128)			

Calculation

The degree of dissociation (α) of acetic acid at any dilution can be calculated using Equation (8.23).

Once α is known, we can calculate the dissociation constant at every dilution using Equation (8.21).

The value of K comes out to be constant and hence proves the validity of Ostwald dilution law.

Result

The dissociation constant of acetic acid is .... and Ostwald dilution law is verified.

POTENTIOMETRIC MEASUREMENT

General Discussion and Instrumentation

Electrode Potential

When a metal is immersed in a solution containing the same metal ion, the following equilibrium is formed: M^{n+}+ne ⇄ M and for this reason, a potential difference is established between the metal and the solution. The potential difference, E can be expressed as:

$$E = E^0 + \frac{RT}{nF}\ln(a_{M^{n+}}) \quad \textbf{(8.24)}$$

Here, the parameters and constants imply as E electrode potential, E^0, standard electrode potential, vary for different metals; both E and E^0 are expressed in terms of volts. R, universal molar gas constant; T, temperature; n, mole fraction; F is Faraday constant, $a_{M^{n+}}$ is activity of M^{n+} ion in solution.

Since ln10=2.303, thus, $\ln(a_{M^{n+}}) = 2.303\,\log_{10}^{(a_{M^{n+}})}$ Put the values of 2.303, R and F, Equation (8.24) becomes:

$$E = E^0 + \frac{0.0591}{n}\log_{10}^{(a_{M^{n+}})} \quad \textbf{(8.25)}$$

For dilute solutions activity is replaced by concentration (in moles/lit). Thus, at 25°C Equation (8.25) can be written as:

$$E = E^0 + \frac{0.0591}{n}\log_{10}^{(c_{M^{n+}})} \quad \textbf{(8.26)}$$

Equation (8.26) is known as the Nernst equation.

To determine the potential difference between an electrode and a solution, it is necessary to have another electrode and solution of accurately known potential difference. The two electrodes can then be combined to form a Voltaic cell. Using the Voltaic cell, the electro motive force (emf) which is nothing but the difference of electrode potential at zero current, can be measured directly.

Using a potentiometer, we can determine the emf of a cell, but not the potential of an electrode.

Formal Potential

In Equation (8.26), the E^0 is measured with respect to the unit activity and with all ions expressed in simple forms, but if the experimental solution is concentrated, then the activity is not equivalent to concentration and chance of formation of complex ion will also increase. Under such conditions activity of a species is smaller than their concentration. Actual species involve may often differ from those to which standard electrode potentials refer. Thus, we can define the formal potential as the standard electrode potentials which are determined under experimental conditions and are denoted by $E^{0'}$. The concentration of the solution, pH of medium, complexation, *etc.* influences the formal potential.

Reference Electrode

A reference electrode is that which has an accurately known electrode potential at 298 K and is used to find the single electrode potentials of other electrodes. The choice of a particular reference electrode depends up on the nature of the solution used. It is often desirable to choose an electrode that will not require a salt bridge. Reference electrodes are categorized as:

1) Primary electrode.

2) Secondary or subsidiary reference electrode.

1) Primary Reference Electrode (Standard Hydrogen Electrode, (SHE)): Standard hydrogen electrode, whose potential is arbitrarily fixed at 0.0 V at 298 K, is employed as a primary reference electrode. Different types of hydrogen electrodes are available but the simple and most common type of hydrogen electrode is due to the Hilderbrand type. It consists of a glass tube, in the lower end of which Pt wire with a Pt foil, coated with Pt black is used. This Pt-foil is surrounded by hydrogen gas at 1 atm pressure (activity=1) and is dipped into a solution containing H^+ ions, say HCl of unit activity ($a_{H^+} a_{H^+} = 1$). For all practical purposes, the unit activity of H_2 gas is taken as an approximation to be 1 bar pressure at the unit activity of H^+ ion is replaced by 1.0 (M) concentration.

Thus, the standard hydrogen electrode is represented as:

$$H^+(aq.a = 1)|H_2(g, a = 1)|Pt$$

$$\text{Or, } H^+(1.0\ M)|H_2(1\ bar)|Pt$$

The hydrogen gas is passed through the side inlet (Fig. **8.26**) and bubbled into the acid solution. After 10-15 min the electrode assumes the constant potential independent of the rate of bubbling. The Pt foil adsorbs a part of hydrogen and the rest comes out through the holes. Thus, equilibrium between adsorbed hydrogen on the electrode surface and ions in the solution is established.

$$H_2(g) \leftrightharpoons 2\,H^+\,(aq.) + 2e.$$

By convention, the potential of this standard hydrogen electrode is arbitrarily fixed at zero. A standard hydrogen electrode is used as a primary reference electrode to determine the relative values of electrode potential (reduction) of other electrodes as there is no method to find out the absolute value of single electrode potential.

2) Secondary Reference Electrode: The use of SHE has several difficulties like setting up and maintenance of unit activity of H^+ ions in solution and pressure of hydrogen gas as 1 atm pressure. It cannot be used in presence of air, oxygen, oxidizing, or reducing agents (safety issue). The Pt-black coating on Pt-electrode can easily be poisoned because of impurities and equilibrium between H_2 (g) and ions in solution are disturbed. To ease the method several secondary or subsidiary reference electrodes with accurately known single electrode potential with respect to hydrogen electrodes are used. Some examples of such types of electrodes are:

(A) Calomel electrode

(B) Quinhydrone electrode

(C) Ag-AgCl electrode

(D) Glass electrode

(A) Calomel Electrode: A calomel electrode consists of a glass tube having a side at each end. A small quantity of pure Hg is placed at the bottom of the vessel, which is covered with a calomel paste of Hg_2Cl_2, Hg, and KCl (all are extra pure). A schematic diagram of the calomel electrode is shown in Fig. (**8.2**). The calomel paste is made by rubbing mercury and calomel (Hg_2Cl_2) in a glass mortar with a small quantity of saturated KCl solution. The paste is quantitatively washed (washed several times) by saturated KCl solution and the mixture is allowed to stand until the calomel has settled and the supernatant part is decanted off.

The vessel is now fitted with saturated KCl solution, by sucking it through the bent tube fitted with a rubber tube and the rubber tube is closed by folding and fitting a pinchcock. The concentration of KCl solution may be of different strength. A glass tube with a Pt-wire sealed into the lower end is fitted into the electrode vessel so that the Pt wire dips in the Hg layer. Pt wire is used to complete the circuit.

The reaction involves in the calomel electrode is:

$Hg\ (l)|Hg_2Cl_2(s)|$ KCl For which the half-cell reaction is:

$$Hg_2Cl_2(s) + 2e \leftrightharpoons 2Hg\ (l) + 2Cl^-\ (aq.)$$

Applying Nernst's equation for e.m.f. we have:

$$E_{cal} = E^0 - \frac{RT}{F} \ln(a_{Cl^-}) \quad \textbf{(8.27)}$$

Thus, the potential of a calomel electrode depends on the activity of Cl^- ions, *i.e.*, the concentration of KCl solution used for the preparation of the electrode. The potentials on hydrogen scale of three reference calomel electrodes are given below:

$Hg\ (l)|\ Hg_2Cl_2\ (s)|$ 0.1 (M) KCl (s); E_{cal}=0.3338+0.00007*(t-25) V

$Hg\ (l)|\ Hg_2Cl_2\ (s)|$ 1.0 (M) KCl (s); E_{cal}=0.28+0.00024*(t-25) V

$Hg\ (l)|\ Hg_2Cl_2\ (s)|$ saturated KCl (s); E_{cal}=0.2415+0.00076*(t-25) V

E_{cal} stands for reduction potential of calomel electrode at t° C. Since the ease of replacing the solution, saturated calomel electrode (SCE) is widely used and commercially more available. However, for most accurate work, a deci-molar (0.1 (M)) calomel electrode is used, because of the minimum temperature coefficient.

To preserve a calomel electrode, it is recommended to wash the outer part of the electrode thoroughly with distilled water after every use and dip the electrode in saturated KCl solution. Do not keep the electrode in distilled water; the concentration of KCl solution inside the electrode will change.

Quinhydrone Electrode: This electrode consists of an inert metal (Pt, Au) dipped into a solution of unknown pH which is saturated with quinhydrone. In an aqueous medium, quinhydrone undergoes the following equilibrium.

$C_{12}H_{10}O_4$	$\leftrightharpoons$	$C_6H_4O_2$ +	$C_6H_6O_2$
$C_6H_4O_2.C_6H_6O_2$		Quinone (Q)	Hydroquinone (H_2Q)
2 Quinhydrone			

Since the aqueous solution of quinhydrone of pH < 7 contains equimolar amounts of quinone and hydroquinone, thus, $c_Q = c_{H_2Q}c_Q = c_{H_2Q}$. Since the solution is dilute, concentration can be replaced by activity. Thus, $a_Q = a_{H_2Q}$.

Since, H_2Q is a weak acid, in the aqueous phase the following reversible oxidation-reduction equilibrium was established: $H_2Q + \leftrightharpoons Q + 2H^+ + 2e$ (electron release, means oxidation)

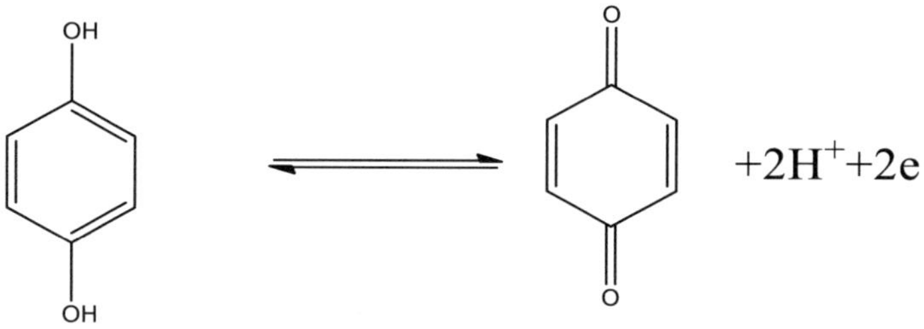

The opposite reaction is $Q+2H^++2e \rightleftarrows H_2Q$ (electron accept, implies reduction). Since the equilibrium involves electron transfer, thus, if a clean Pt-electrode is dipped into the solution, then potential will be developed and the potential can be expressed as:

$$E = E^0 + \frac{RT}{F} \ln\left(\frac{a_Q.a_{H^+}^2}{a_{H_2Q}}\right)$$

$$= E^0 + \frac{RT}{F} \ln (a_{H^+})$$

$$E^0 + \frac{RT}{F} \ln(a_{H^+})$$

(Since the activity of other terms are same)

$$= E^0 + 0.05911 \log(a_{H^+}) \text{ at } 25° \text{ C } (298.15\text{K})$$

$$=E^0 - 0.05911\text{pH} \qquad \textbf{(8.28)}$$

E^0 for this system is 0.6994 V at 25° C. Thus, $E = 0.6994 - 0.05911\text{pH}$ at 25° C (298.15 K). This shows that emf. of the quinhydrone electrode is pH-dependent or in other words, we can say these electrodes behave as reversible hydrogen electrodes.

Setting a quinhydrone electrode is done by adding a pinch of quinhydrone to the experimental solution (pH to be determined). The quinhydrone attains equilibrium very quickly and then dips a Pt-electrode in solution. Thus, it is advantageous to too easy to set. Quinhydrone electrode can give accurate pH values up to pH 8. Above pH 8, the medium contains an excessive number of hydroxyl ions, resulting in quinone is attacked by the hydroxyl group and the equilibrium is disturbed. Thus, $\frac{a_Q}{a_{H_2Q}} \neq 1$. For solution having pH>8, quinhydrone electrode used as anode with respect to SCE and potentiometer shows negative emf values.

Silver/silver Chloride Electrode

For accurate measurements of chloride solutions, Ag-AgCl electrodes are widely used to bypass the use of liquid junction potential. Generally, this type of electrode consists of Ag wire/Ag coated Pt wire, dipping into a solution of KCl or HCl of known concentration which is saturated with AgCl (Prepared by adding 2-3 drops of 0.1 (M) $AgNO_3$ solution and shake).

Representation of this electrode: Ag/AgCl (s)/Cl^- (aq)

The electrode involves following reactions: $AgCl\,(s) + e \rightleftarrows Ag(s) + Cl^-$ and the electrode potential:

$$E = E^0 - \frac{RT}{F}\ln(a_{Cl^-}) \qquad \textbf{(8.29)}$$

Like the calomel electrode, the Ag-AgCl electrode is also a reversible electrode with respect to the anion (Cl^-), thus, at a constant temperature, the only variable on the right-hand side is the activity of chloride ions from the KCl solution. At NTP (25° C, 1 atm pressure), the standard reduction potential for this system is 0.223 V.

Glass Electrode

The glass electrode is the most widely used H^+ ion responsive electrode and its use is dependent upon the fact that when glass membrane is immersed in a solution containing H^+ ion. In other words, potentially becomes pH-dependent.

The glass electrode consists of a very thin-walled glass tube and low melting, highly conducting glass is used. This is blown at the end of the glass tube. A solution of 0.1 (M) HCl which furnishes a constant H^+ ion is kept inside the bulb. An Ag-AgCl or calomel electrode is immersed into it to make electrical contact.

Several types of glass electrodes are available. Some of them contain an indicator electrode (thin glass bulb) and a reference electrode (Ag-AgCl electrode) combined in a single unit. The thin glass bulb and the narrow tube to which it is attached are filled with 0.1 (M) HCl and carry an Ag-AgCl electrode. The wide tube is fused to the lower end of the narrow tube and contains saturated KCl solution which is also saturated with AgCl. The tube contains saturated KCl solution, also carries an Ag-AgCl electrode. The whole system is encapsulated by an insulating cap.

Commercial glass electrodes come in combination with Ag-AgCl electrode/calomel electrode. For the first use of glass electrodes, overnight immersion of the electrode in distilled water is advisable, such that a thin layer of water will be formed and ion exchange can take place. Suppose the glass of a glass electrode contains Na^+ ion near the surface, the glass is replaced by H^+ ion and equilibrium between Na^+ ion and H^+ ion, which can be represented as: $H^+_{solution}+Na^+_{glass} \rightleftarrows H^+_{glass}+Na^+_{solution}$. The exchange of charge produces a potential difference between the glass and the solution and this difference in potential is the potential of the glass electrode which is pH-dependent. The electromotive force can be expressed as: $E_G=E^0_G-0.05911pH$ at 25° C. E^0_G is unknown and partially depends on the membrane composition, the character of the electrode. Thus, E^0_G has to be measured for every case. Like standard hydrogen electrodes, glass electrodes behave reversibly with H^+ ions in combination with a suitable electrode. Thus, they are also useful to measure pH. The cell can be represented as:

Glass electrode| test solution||KCl, Hg_2Cl_2 (s) (SCE)|Hg

Thus, Ag|AgCl (s), 0.1 (M) HCl| Glass| test solution|| KCl, Hg_2Cl_2 (s) (SCE)|Hg

E^0_G can be measured by calibrating emf of a buffer solution of known pH.

For the above cell: $E = E_{cal} - E_G = 0.2415 - E^0_G + 0.059$ pH at 25° C **(8.30)**

The solution of unknown pH is used and the pH of that solution is ascertained. Since glass is a high resistance material, it cannot be used in potentiometry; instead, it can be used for electronic highly digital voltmeters, like pH.

Convention of Representation of Cells and Calculation of Cell Electro-Motive Potential

Any cell consists of two half-cells, which are nothing but two electrodes. Between these two electrodes, an electrode with a higher standard reduction potential value should be placed on the right-hand side electrode (RHE) and considered as the positive electrode. Automatically, another electrode plays the role of the negative electrode and is placed on the left-hand side of the electrode (LHE). These two electrodes are connected *via* a salt bridge to eliminate liquid junction potential and are indicated by a couple of vertical lines (||). Thus, a complete cell can be represented as:

$$\underset{\text{Left-handelectrode(LHE)}}{(-)} \,||\, \underset{\text{Right-handelectrode(RHE)}}{(+)} \quad \textbf{(8.31)}$$

Since If E^0_R and E^0_L are standard reduction potential of RHE and LHE. From the above discussion, $E^0_R > E^0_L$. If E_R and E_L represent are the electrode potential (in reduction form) of RHE and LHE, then the observed emf of the cell (E_{cell}) is $E_{cell} = E_R - E_L$. E_R and E_L can be evaluated from the Nernst equation.

Measurement of Cell EMF

The EMF of a cell is the potential difference responsible for the flow of current from one electrode of higher potential to the other of lower potential. Unit of EMF is volt.

EMF is measured by the principle that the emf of the cell is just balanced by an equal and opposite difference so that no net current flows. This is achieved by balancing the unknown emf against a known potential drop, which can be varied, using the following circuit.

Batteries are nothing but storage cell (~2 V) is the steady source of emf (large cell) greater than the EMF to be measured, which sends a current in the calibrated resistance AB with a movable contact X. AB is potentiometer wire. Generally, this consists of Pt or Pt-Ir alloy. If no current passes through X, then there is a uniform potential gradient along with AB and the potential difference across any section on it depends upon the resistance of that part.

If cell C_1 of emf is E_x is joined through a galvanometer G to the sliding contact X in AB and point X is adjusted so that there is no flow of current which is reflected by no deflection of galvanometer indicator. In this stage potential drop along AX due to battery is exactly compensated by emf of C_1.

It is advisable to use a standard cell at C_2 of exactly known emf. (E_x). By using a suitable switch, cell C_1 is replaced by standard cell C_2 and the point of balance X′ is noted. Here, the fall of potential along AX′ = E_s. Thus,

$$\frac{\text{emf of } C_1}{\text{emf of } C_2} = \frac{E_X}{E_S} = \frac{\text{Resistance of AX}}{\text{Resistance of AX}'} \quad \textbf{(8.32)}$$

Equation (8.32) can be used for the calculation of E_x. However, during measurement of E_x is redundant. In the earlier manual potentiometer, a scale (in volt unit) is fitted in connection with sliding contact which is a movable knob. After balancing E_s, E_x is connected to the potentiometer and the galvanometer shows deflection. The movable knob is then adjusted at the balance point and E_x can be measured from the scale.

Standard Cell

A standard cell of known emf is not used to get current supply, instead, they are used for comparison of emf.

A cell is considered a standard cell if it is capable of giving constant and reproducible emf independent of change of temperature.

The most convenient and generally used standard cell is the Weston cell. In this cell, Hg forms a positive pole and a 12.5% Cd-Hg (cadmium amalgam) is the negative pole. A saturated solution of cadmium sulphate is used ($CdSO_4$, $8H_2O$) as an electrolyte. The open ends of the arms are covered by wax to prevent the evaporation of water. The following reactions take place in a Weston cell:

$$Cd + Hg_2^{2+} \leftrightharpoons Cd^{2+} + 2Hg$$

The positive pole is generally covered up with a paste of sparingly soluble Hg_2SO_4 and $CdSO_4$ in Hg. The emf of the cell is 1.0183 V at 20° C and change of electrode potential (E) at a particular temperature, t° C can be corrected as E = 1.0183 – 4.06 × 10^{-5} × (t-20). The cell can be represented as:

$$12.5\%\ \text{Cd in Hg}\ |\ CdSO_4, \frac{3}{8} H_2O(s)\text{saturated solution}: Hg_2SO_4|Hg$$

12.5% Cd in Hg| $CdSO_4$, 3/8 H_2O (s) saturated solution: Hg_2SO_4|Hg. To get constant EMF, it is advisable not to keep the cell in the circuit for a long time.

Salt Bridge

A salt bridge is generally a double bend glass tube, containing a viscous gel of saturated solution of agar-agar and a strong electrolyte, whose ionic (corresponding ions) mobility are of same ionic mobility or almost equal transport number in aqueous solution and not reached with the two electrolytic solutions of the cell.

Generally, KCl, KNO_3, NH_4NO_3, *etc.* are used as an electrolyte in the preparation of salt bridge (transport no. of K^+, Cl^-, NH_4^+, NO_3^- are 7.62, 7.92, 7.6, 7.4 $m^2 V^{-1} sec^{-1} \times 10^8$).

Salt Bridge Preparation

Take ~60 ml of distilled water in a 250 ml beaker. Prepare a saturated solution of either of the above-mentioned electrolytes by heating. Add ~2 g of agar-agar powder in hot condition into it. Again, boil the mixture gently and stir on a steam bath till a clear solution appears. Pour the solution carefully into the double bent glass tube thoroughly. While filling, please ensure that no bubble got trapped inside the tube. Dip the tube into an ice bath (ensure zero water contamination from the open end of the tube, it can damage the gel character) and the gel becomes solidified.

Role of A Salt Bridge

(1) It connects the electrodes, *i.e.*, using a salt bridge, the inner circuit of a cell (unknown emf) gets completed.

(2) Liquid junction potential which occurs at the interface of electrolyte and salt solutions is eliminated or minimized.

For acid-base titration KCl, KNO_3, NH_4NO_3 can be used, but if the solution contains Cl^- or halide ions, the KCl salt bridge can never be used, but the rest two can be used.

Calibrating of a Digital Potentiometer

In the present era of digitalization, a digital potentiometer is widely available. In the digital potentiometer, Poggendorff's compensation method of measurement of emf of a cell (against a standard electrode) is taken care of by modifying the electronic circuit. These differ in various potentiometers, but the following two general methods are widely followed in most of the potentiometers:

(1) The instrument is working in the 220-230 V electric line. Connect the electrodes of the cell to the meter through a jack and then set the switch-in/switch-out meter. Switch on the power supply. The display of the meter will show 1.018 V, if not, then adjust it to 1.018 V by slightly moving the "adjust" / "calibrate" knob. Once the adjustment is over, set the switch-in/switch-out of the meter and 0.0000 V will be displayed if the electrode is immersed into the distilled water. If not note the reading, subtract this value from the next successive readings. Now the instrument is ready to measure emf of given unknown cell. Just connect the cell and note the displayed readings.

(2) To calibrate another set of instruments, ask for a buffer solution of pH7. pH 7 buffer capsules are readily available in the market. Dissolve the buffer capsule completely in 100 ml distilled water. Connect the glass electrode and an SCE to the potentiometer and switch it on. Immerse the electrode assembly in the buffer solution and 0.000 ±0.001 V will be displayed on the display board and the potentiometer is ready to measure EMF of the given unknown cell. If the value is not displayed, then adjust as described above to 0.000 ±0.001 V.

Within these two types of the potentiometer, some instruments provide a scale expansion facility, where the scaling can be increased to higher digits. This gives more accuracy.

It is to be noted that we have used a digital potentiometer. So, from here onwards we use the potentiometer term, instead of the digital potentiometer.

I. POTENTIOMETRIC TITRATION OF A SOLUTION OF A STRONG ACID WITH A SOLUTION OF STRONG BASE USING QUINHYDRONE ELECTRODE

Chemicals and Apparatus

Quinhydrone, 0.01 (M) HCl and NaOH, 0.02 (M) oxalic acid solution, potentiometer, Pt electrode, saturated calomel electrode, KNO_3 salt bridge, burette, pipette, beaker (4 pieces), volumetric flask (100 ml, 2 pieces).

Theory

Quinhydrone is composed of an equimolar mixture of quinone (Q) and hydroquinone (H_2Q). In the aqueous phase, it attains the following equilibrium:

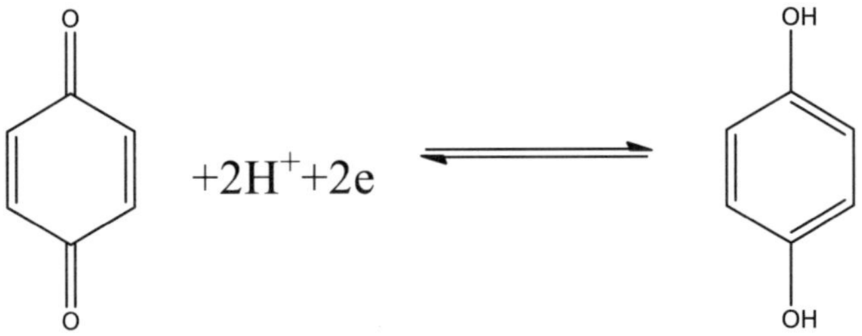

In short, we can write: $Q + 2H^+ + 2e \leftrightharpoons H_2Q$, which involves 2 electron transfer processes. Thus, if a clear Pt-electrode is dipped into an acidic solution of quinhydrone, it can play the role of an electrode of the type: quinhydrone H^+|(Pt). The reduction potential of quinhydrone can be represented by Nernst's equation as:

$$E=E_Q^0 + \frac{RT}{2F}\ln\left(\frac{a_Q a_{H^+}^2}{a_{H_2Q}}\right) \quad \textbf{(8.33)}$$

'a' stands for activity. Since Q and H_2Q are in solid-state, so we can consider their activity=1. Thus, the above equation can be rewritten as:

$$E=E_Q^0 + \frac{RT}{F}\ln(a_{H^+}) \quad \textbf{(8.34)}$$

If this electrode is coupled with a standard SCE, then the cell can be expressed as:

$$\overset{(-)}{\text{SCE(LHE)}}||\text{solution of } H^+, \overset{(+)}{\text{Quinhydrone(RHE)}}|\text{Pt}$$

The observed EMF of the cell will be:

$$E_{cell}=E_L - E_R=E_R^0 + \frac{RT}{F}\ln(a_{H^+})\text{-}E_{cal} \quad \textbf{(8.35)}$$

Here E_{cal} is the standard reduction potential of SCE. The E_L is nothing but the standard reduction potential of quinhydrone (E_Q^0). At 25° C, Equation (8.35) turns out to be:

$$E_{cell} = E_Q^0 - E_{cal} + 0.05911 \log(a_{H^+}) \quad \textbf{(8.36)}$$

Equation (8.36) is the working equation. From Equation (8.36) it is evident that after the complete set of a cell if alkali is added to an acid solution (saturated with quinhydrone), the pH of the solution will increase, and thus, E_{cell}will decrease. At equivalence point a sharp decrease in E_{cell} will be observed, which indicates the equivalence point. Once the equivalence point is achieved, further addition of base will give negative EMF of the cell, indicates a change in polarity.

Distribution of the experiment can be organized as first standardization of NaOH solution by standard oxalic acid solution by adopting potentiometric titration method and then find out the strength of HCl using standardised NaOH solution using the same procedure. In the first case, the plot of change of potential (ΔE) for the addition of n numbers of drops/volume of NaOH solution (ΔE/n) *vs.* n will be as Fig. (**8.8**). In this plot, a double-humped graph is observed, an indication of the dibasic nature of oxalic acid. The peak of the second hump indicates the equivalence point of the titration. From the number of drops volume of NaOH can be calculated and hence the strength of NaOH also. On the other hand, the same for the titration of HCl *vs.* NaOH, the plot will be like Fig. (**8.9**). The X-axis corresponding to the peak of the graph represents the volume of NaOH of known strength required to titrate a certain amount of HCl of unknown strength. So, from this titration, we can get the exact strength of HCl because the volume and strength of NaOH are known and the volume of HCl is also known.

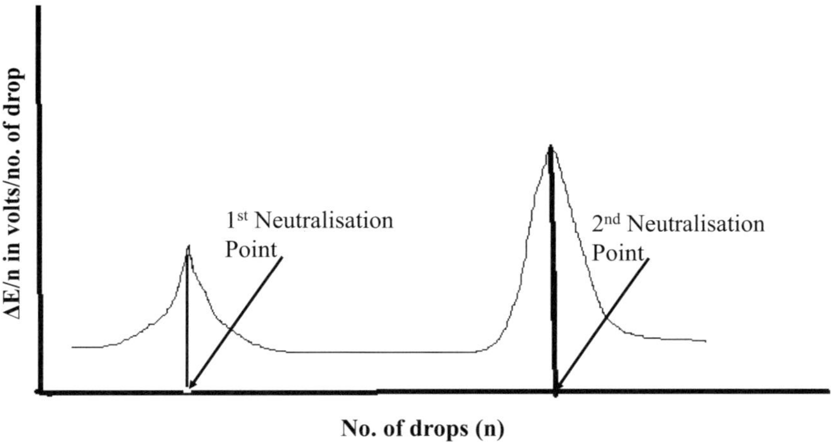

Fig. (8.8). Potentiometric titration of oxalic acid solution *vs.* NaOH solution.

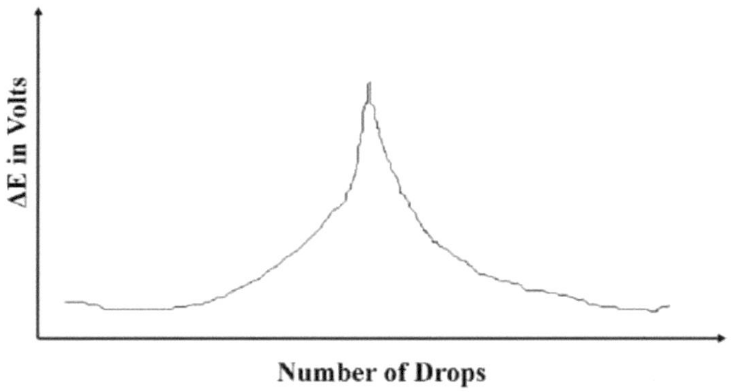

Fig. (8.9). Potentiometric titration of HCl solution *vs.* NaOH solution.

Calculation of Strength of NaOH Solution

Count accurately 1ml = x no. of drops. Therefore, the volume of 1 drop = 1/x

Strength of oxalic acid solution = S_1 (M)

Volume of oxalic acid taken = V_1 ml

No. of drops of NaOH required to titrate the V_1 ml of S_1 (M) oxalic acid = n_1

Therefore, the volume of that consumed NaOH consumed = n_1/x ml

Thus, strength of NaOH (S_2) = [($V_1 \times S_1$)/(n_1/x)] (M)

Calculation of Strength of HCl Solution

The volume of HCl solution of unknown strength taken = V_2ml.

No. drops of NaOH (strength: S_2 (M)) are required to titrate the V_2 ml HCl = n_2.

Therefore, the volume of that consumed NaOH consumed = n_2/x ml.

Thus, strength of HCl (S_3) = [((n1/x) × S_2)/ V_2] (M).

To get a sharp peak of the first titration, it is advisable to take quinhydrone as minimum as possible.

Note: Students can measure volume directly instead of counting number of drops.

Procedure

1) Prepare 100 ml of 0.01 (M) oxalic acid solutions (Table **8.22**) by accurate weighing also prepare 100 ml of NaOH solution of an approximate strength of 0.01 (M).

2) Pipette out 10 ml of oxalic acid solution and keep it in a 100 ml beaker. Add a pinch of quinhydrone to it and make a uniform solution by gentle shaking. Dip a clean Pt-electrode into the solution and connect the quinhydrone electrode with the SCE through a KNO_3-Agar salt bridge. Note the zero reading, *i.e*, without adding NaOH.

3) Add 1 drop of NaOH solution by using a burette and note the potentiometer reading. Continue the same at least 5 to 6 negative reading appears.

4) Plot $\Delta E/n$ *vs.* n and from the second peak, calculate the strength of NaOH (Table 8.23).

5) Wash the electrode, beaker, outer wall of the salt bridge thoroughly and take 10 ml of HCl solution of unknown strength. Repeat steps 2-4. From the peak of the plot, the titration point can be measured accurately.

Calculation and Results

Temperature = ... ° C

Preparation of 100 ml ~0.1 (M) NaOH solution: Dissolve 0.4 g NaOH beads in 100 ml water in a beaker.

Table 8.22. Preparation of 100 ml 0.01 (M) oxalic acid solution.

Molecular weight of oxalic acid ($C_2O_4H_2.2H_2O$=126)

Mass of Empty Weighing Bottle (w_1) (g)	Mass of Weighing Bottle + Oxalic Acid (w_2) (g)	Mass of Oxalic Acid Transferred (w_1 - w_2) (g)	Mass of Oxalic Acid to be Taken (g)	Strength of Oxalic Acid Solution (M)
			0.126	

Table 8.23. Standardisation of NaOH solution by a standard oxalic acid solution using potentiometric titration method.

Volume of oxalic acid taken= 10 ml.

S. No.	No. of Drops of NaOH Added (n)	Observed EMF (V or mV)	ΔE/Δn (V/drop)
1	0		
2	1		
3	3		
4	4		

Table 8.24. Titration of HCl solution by standardised NaOH solution using potentiometric titration method.

Volume of HCl taken = 10 ml.

S. No.	No. of Drops of NaOH Added (n)	Observed EMF (V or mV)	ΔE/Δn (V/drop)
1	0		
2	1		
3	3		
4	4		
.	.		

Calculation of Strength of HCl Solution

From graph:

The volume of NaOH required for oxalic acid = ... ml.

So, the strength of NaOH solution = ... (M).

The volume of NaOH required for HCl solution = ... ml.

So, the strength of HCl solution = ... (M).

II. TO FIND OUT THE STRENGTH OF GIVEN MOHR'S SALT SOLUTION BY TITRATING IT WITH 0.1 (N) $KMnO_4$ SOLUTION POTETIOMETRICALLY. ALSO FIND THE REDOX POTENTIAL OF FERROUS-FERRIC SYSTEM

Chemicals and Apparatus

Mohr's salt solution in 2(N) H_2SO_4, potassium dichromate solution, Pt-electrode, SCE, KCl-Agar salt bridge, potentiometer, beaker, burette, pipette.

Theory

In aqueous medium, ferrous and ferric ions assume the following equilibrium:

$$Fe^{3+} + e \rightleftharpoons Fe^{2+}$$

Nernst's equation can be expressed as:

$$E = E^{0}_{Fe^{3+}/Fe^{2+}} + \frac{RT}{F} \ln \left(\frac{c_{Fe^{3+}}}{Fe^{2+}} \right) \quad \textbf{(8.37)}$$

When a Pt-wire is dipped into the system, it will serve as an electrode of the type: Fe^{3+}- Fe^{2+}/ (Pt). When this electrode is coupled with a reference electrode such as SCE and these two electrodes are connected through a KCl-Agar salt bridge, the cell set up will be:

$$\overbrace{\text{SCE}}^{-} \; || \; \overbrace{Fe^{3+} - Fe^{2+}}^{+} \; | \; \text{Pt} \quad \overset{(-)}{\text{SCE}} || \overset{(+)}{Fe^{3+}} - Fe^{2+} | \text{Pt} \quad \textbf{(8.38)}$$

The emf of the corresponding cell will be:

$$E_{cell} = E_R - E_L = E^0_{Fe^{3+}/Fe^{2+}} + \frac{RT}{F}\ln\left(\frac{c_{Fe^{2+}}}{c_{Fe^{3+}}}\right) - E_{cal} \quad \textbf{(8.39)}$$

E_{cal} represents the formal (reduction) potential of SCE and $E^0_{Fe^{3+}/Fe^{2+}}$ is the standard reduction potential of the ferrous-ferric system. At 25° C Equation (8.39) can be written as:

$$E_{cell} = E^0_{Fe^{3+}/Fe^{2+}} - E_{cal} + 0.0591\log\left(\frac{c_{Fe^{2+}}}{c_{Fe^{3+}}}\right) \quad \textbf{(8.40)}$$

If $K_2Cr_2O_7$ (dichromate) solution is added to the system, Fe^{2+} ion is removed and the concentration of Fe^{3+} increases. With the progressive addition of dichromate solution, the observed emf (E_{cell}) will gradually increase. Near the equivalence point, almost all ferrous ions get oxidized and a sharp change will be observed in that region. Just after the equivalence point, instead of Fe^{2+}/Fe^{3+} equilibria, $Cr_2O_7^{2-}/Cr^{3+}$ predominates. The second one has more standard reduction potential compared to the first one. Hence, a sharp increase in emf will be observed

From Equation (8.40), at half-equivalence point $c_{Fe^{2+}} = c_{Fe^{3+}}$;

$$(E_{cell})_{1/2} = E^0_{Fe^{3+}/Fe^{2+}} - E_{cal} \quad \textbf{(8.41)}$$

Thus,

$$E^0_{Fe^{3+}/Fe^{2+}} = E_{cal} + (E_{cell})_{1/2} \quad \textbf{(8.42)}$$

Here, $E^0_{Fe^{3+}/Fe^{2+}}$ is the formal reduction potential

Thus, the plot of E_{cell} *vs.* no. of $K_2Cr_2O_7$ solution (n) or volume added, the following sigmoid curve (Fig. **8.10**) will be obtained. Since E_{cal} is known with temperature correction, so from the graph $(E_{cell})_{1/2}$ can be obtained and $E^0_{Fe^{3+}/Fe^{2+}}$ can also be obtained.

A plot of $\Delta E_{cell}/\Delta n$ *vs.* n or volume (Fig. **8.11**) gives the exact volume of $K_2Cr_2O_7$ required for neutralisation. Since the volume and strength of $K_2Cr_2O_7$ are known and the volume of Mohr's salt is known, thus, the strength of Mohr's salt solution can be calculated easily.

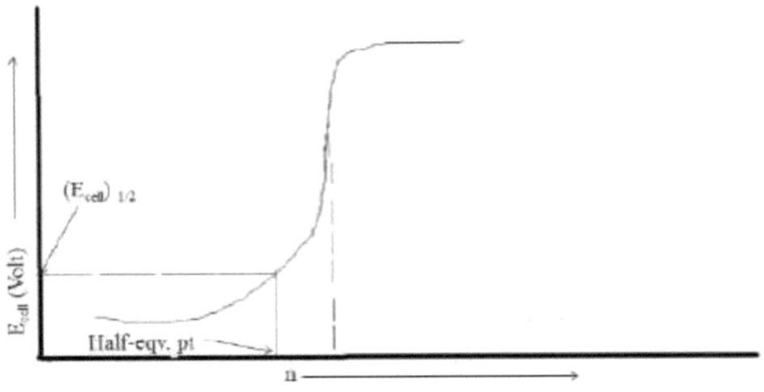

Fig. (8.10). Potentiometric titration of Mohr salt solution by standard $K_2Cr_2O_7$ solution. The plot of E_{cell} *vs.* no. of drops of $K_2Cr_2O_7$.

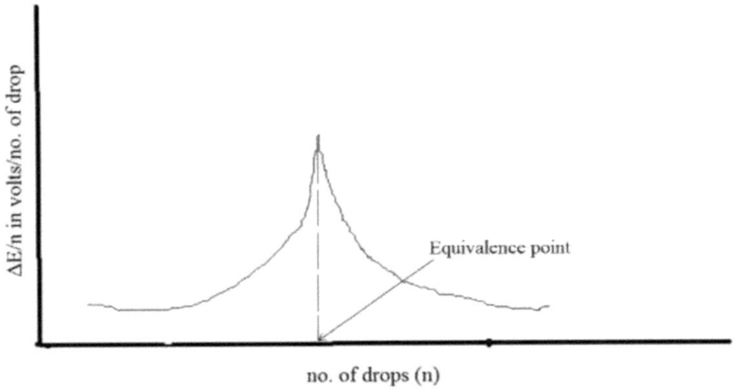

Fig. (8.11). Plot of $\Delta E_{cell}/\Delta n$ *vs.* n for added (n) for potentiometric titration of Mohr salt.

Procedure

1) Prepare 100 ml of each of ~0.05 (M) Mohr salt solution in 0.05 (M) H_2SO_4 and 0.5 (M) $K_2Cr_2O_7$ solution (Table **8.25**).

2) Pipette out 25 ml Mohr salt solution in a 100 ml clean beaker. Insert the electrodes into the solution. Calibrate the potentiometer to the reading mode and take the initial reading of the meter before the addition of $K_2Cr_2O_7$ solution into the Mohr solution.

3) Add 2 drops of $K_2Cr_2O_7$ solution-shake gently and take the meter reading. Continue addition and note meter reading each time (Table **8.26**). Take 5 to 6 more readings beyond the equivalence point (after the sharp change in emf and a persistent green solution will remain in the beaker).

4) Clean the electrode thoroughly with distilled water and keep immersed into distilled water.

5) Plot a couple of graphs; 1) between E_{cell} *vs.* n and Δ $E_{cell}/\Delta n$ *vs.* n. From the graph calculate the strength of the Mohr solution and the formal potential of the ferrous-ferric system. Plot the same by using volume instead of number of drops.

Observations and Results

Temperature = ...° C

Table 8.25. Preparation of 0.05 (M) $K_2Cr_2O_7$ solution.

Molecular weight of $K_2Cr_2O_7$ solution = 249.22

Mass of empty Weighing Bottle (w1) (g)	Mass of Weighing Bottle + $K_2Cr_2O_7$ (w2) (g)	Mass of $K_2Cr_2O_7$ Transferred (w1 - w2) (g)	Mass of $K_2Cr_2O_7$ to be Taken (g)	Strength of $K_2Cr_2O_7$ Solution (M)
			1.2461	

The strength of commercially available H_2SO_4 is 36 (N) or 18 (M). So, take 5.6 ml of the same and dilute to 100 ml in cold condition to prepare 1 (M) H_2SO_4 solution.

The molecular weight of Mohr's salt ($FeSO_4$, $(NH_4)_2$ SO_4. $6H_2O$) solution = 392.16.

Take ~1.96 g Mohr's salt and add it to the 1 (M) H_2SO_4 solution.

Table 8.26. Titration and recording of emf.

Volume of Mohr's salt solution taken = 25 ml.

Sl. No.	No. of Drops of $K_2Cr_2O_7$ Added (n)	Observed EMF (V or mV)	$\Delta E/\Delta n$ (V/drop)
1	0		
2	1		
3	3		
4	4		
.	.		

Calculation of Strength of Mohr's salt Solution and Formal Potential of Fe^{3+}-Fe^{2+} System

(1) From plot E_{cell} *vs.* n, $E^0_{Fe^{3+}/Fe^{2+}} = E_{cal} + (E_{cell})_{1/2}$ = ...V. Reported value = +0.68 V in 1(M) H_2SO_4 solution.

(2) From ΔE/Δn *vs.* n, the number of drops of $K_2Cr_2O_7$ solution required = ... ml. Hence strength of Mohr's salt solution=.... (M) (Use: V1S1 = V2S2). Students can measure volume directly instead of counting drops.

III. TO DETERMINE THE STANDARD REDUCTION POTENTIAL OF $[Fe(CN)_6]^{4-}/[Fe(CN)_6]^{3-}$ SYSTEM

Chemicals and Apparatus

0.1(M) solutions of both $K_4[Fe(CN)_6]$ and$K_3[Fe(CN)_6]$, Pt-electrode and SCE, KCl-Agar salt bridge, potentiometer, beaker, burette, pipette.

Theory

Ferrocyanide and ferricyanide ($[Fe(CN)_6]^{4-}/[Fe(CN)_6]^{3-}$) in the aqueous system assumes the following equilibrium:

$$[Fe(CN)_6]^{3-} + e \leftrightharpoons [Fe(CN)_6]^{4-}$$

The system can show the difference if a Pt-electrode is dipped into the solution and the potential can be expressed by using the Nernst equation as:

$$E = E^0 + \frac{RT}{F}\ln\left(\frac{a_{[Fe(CN)_6]^{3-}}}{a_{[Fe(CN)_6]^{4-}}}\right) \quad \textbf{(8.43)}$$

Since activity is nothing but concentration (c) multiplied by activity coefficient (f). Thus, we can write:

$$=E^0 + \frac{RT\times 2.303}{F}\log\left(\frac{c_{[Fe(CN)_6]^{3-}}\times f_{[Fe(CN)_6]^{3-}}}{c_{[Fe(CN)_6]^{4-}}\times f_{[Fe(CN)_6]^{4-}}}\right) + \frac{RT\times 2.303}{F}\log\left(\frac{f_{[Fe(CN)_6]^{3-}}}{f_{[Fe(CN)_6]^{4-}}}\right)$$

$$E = E^0 + \frac{2.303\,RT}{F}\log\left(\frac{c_{[Fe(CN)_6]^{3-}}}{c_{[Fe(CN)_6]^{4-}}}\times\frac{f_{[Fe(CN)_6]^{3-}}}{f_{[Fe(CN)_6]^{4-}}}\right) \quad \textbf{(8.44)}$$

Applying Debye-Hückel limiting law, we can correlate between f and ionic strength (μ) as:

$\log f = -AZ^2$, where A is the Debye-Hückel constant and μ can be expressed in terms of concentration and charge:

$$\mu = \frac{1}{2}\sum_i c_i z_i^2 \quad \textbf{(8.45)}$$

Thus, the third term in the RHS of Equation (8.44) can be written as:

$$\log\left(\frac{f_{[Fe(CN)_6]^{3-}}}{f_{[Fe(CN)_6]^{4-}}}\right) = -A\left(Z^2_{[Fe(CN)_6]^{3-}} - Z^2_{[Fe(CN)_6]^{4-}}\right)\sqrt{\mu}$$

$$\log\left(\frac{f_{[Fe(CN)_6]^{3-}}}{f_{[Fe(CN)_6]^{4-}}}\right) = -A\left(Z^2_{[Fe(CN)_6]^{3-}} - Z^2_{[Fe(CN)_6]^{4-}}\right)\sqrt{\mu} \quad \textbf{(8.46)}$$

Thus, an equimolar mixture of $K_4[Fe(CN)_6]$ and$K_3[Fe(CN)_6]$, then 2nd term in the RHS of Equation (8.44) becomes unity. Thus, in this situation applying Equation (8.45), Equation (8.44) can be written as:

$$E=E^0 + \frac{RT\times 2.303}{F} A\left(Z^2_{[Fe(CN)_6]^{4-}} - Z^2_{[Fe(CN)_6]^{3-}}\right)\sqrt{\mu} \quad \textbf{(8.47)}$$

Since all the terms except μ in the 2nd term of RHS of Equation (8.47) are constant, so Equation (8.47) can be written as:

$$E=E^0+K\sqrt{\mu} \quad \textbf{(8.48)}$$

$$K = \frac{2.303\ RT}{F} A\left(Z^2_{[Fe(CN)_6]^{3-}} - Z^2_{[Fe(CN)_6]^{4-}}\right) \quad \textbf{(8.49)}$$

Thus, to find out E^0, we can set the following cell set up:

$$\overset{(-)}{SCE} \| \overset{(+)}{[Fe(CN)_6]^{3-}} \text{-} \overset{(+)}{[Fe(CN)_6]^{4-}} /Pt \quad \textbf{(8.50)}$$

The observed EMF of the cell is given by:

$$E_{cell}=E_R-E_L$$

$$=E^0+K\sqrt{\mu}-E_{cal}$$

$$= (E^0\text{-}E_{cal}) + K\sqrt{\mu} \qquad \textbf{(8.51)}$$

Thus, from Equation (8.51), the plot of Ecell *vs.* $\sqrt{\mu}$ for different solutions will be a straight line with an intercept of (E^0-E_{cal}) (Fig. **8.12**). Since, at a particular temperature, E_{cal} can be found using the following formula:

$$E_{cal} = \left[0.2415\text{-}7\times10^{-4} \times (t\text{-}25)\right] \qquad \textbf{(8.52)}$$

Hence, E^0 can easily be calculated.

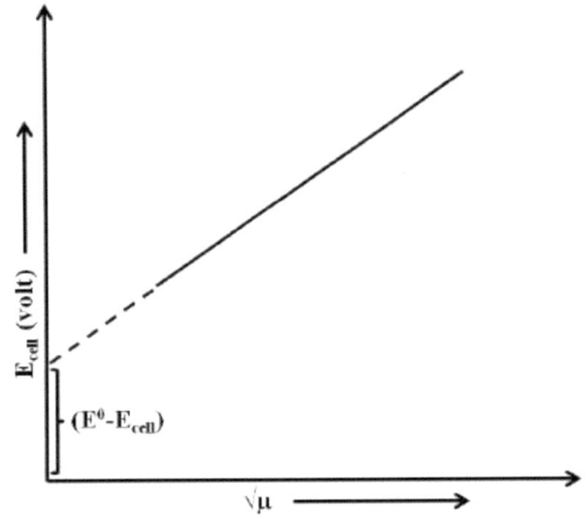

Fig. (8.12). Plot of Ecell *vs.* $\sqrt{\mu}$.

Procedure

1) Prepare exact 100 ml 0.1 (M) $K_4[Fe(CN)_6]$ and $K_3[Fe(CN)_6]$ solution. If not then make exactly 100 ml solution of both by necessary dilution.

2) Take 50 ml 0.1 (M) of both the prepared solutions and mix thoroughly. Prepare the following sets (Table **8.27**) of solutions by accurate dilution:

Table 8.27. Preparation of solution.

100 ml Dry Beaker No.	Volume of 0.05 (M) Mixture (ml)	Volume of Distilled Water (ml)	Concentration of Resultant Mixture (M)
1	4	16	0.010

2	6	14	0.015
3	8	12	0.020
4	10	10	0.025
5	12	8	0.030
6	14	6	0.035

3) Insert a clean Pt-electrode to solution no. 1 and then connect the electrodes through the KCl-Agar salt bridge and measure EMF (Table **8.28**).

4) Repeat 3) for each of the prepared solutions and plot E_{cell} *vs.* $\sqrt{\mu}$ (Table **8.29**) and calculate E° from intercept

Results and Calculation

Temperature =...° C

Table 8.28. Recording of EMF of different sets.

Solution no.	Molar Concentration Equimolar Mixture of $K_4[Fe(CN)_6]$ and $K_3[Fe(CN)_6]$ (M)	Observed emf (E_{cell}) (V)
1	0.010	
2	0.015	
3	0.020	
4	0.025	
5	0.030	
6	0.035	

Table 8.29. Table for graph.

Solution no.	Molar Concentration Equimolar Mixture (M)	$\sqrt{\mu} = \sqrt{\left(\frac{1}{2}\sum c_i Z_i^2\right)}$	E_{cell} (V)
1	0.010		
2	0.015		
3	0.020		
4	0.025		
5	0.030		
6	0.035		

From the graph, intercept =.... V. Thus, E^0= (Intercept + E_{cal}) volt

(Reported value= +0.36 V) at 25° C

V. TO DETERMINE TRANSPORT NUMBERS OF SILVER ION AND NITRATE ION IN SOLUTIONS OF SILVER NITRATE IN THE CONCENTRATION RANGE 0.01 (M) TO 1 (M)

Chemicals and Apparatus

$AgNO_3$ solution, KNO_3 solution, KNO_3-Agar salt bridge, Pt-electrode, SCE, potentiometer.

Theory and Procedure

Construct the following concentration cell:

$$\text{Ag}\left|\underset{(a_1)}{0.01}\text{(M) AgNO}_3\text{ (aq.)}\right|\left|\underset{(a_2)}{0.1}\text{(M) AgNO}_3\text{ (aq.)}\right|\text{Ag}$$

Take two 100 ml beakers half-filled with the respective solutions. Measure the EMF of the cell using bridges of 0.1 (M) $AgNO_3$ (E_1) and saturated KNO_3 (E_2) solutions. The measurements are repeated with fresh solutions to check for the errors due to the diffusion of solute. We have,

$$E_1 = 2t_{NO_3^-}\left(\frac{2.303RT}{F}\right)\log\left(\frac{a_2}{a_1}\right)$$

$$E_2 = \left(\frac{2.303RT}{F}\right)\log\left(\frac{a_2}{a_1}\right)$$

Calculate $t_{NO_3^-}$ using the activity coefficients data from E_1 and compare with the value obtained from the relation: $t_{NO_3^-} = 2\frac{E_1}{2E_2}$

Results

The transport number of Ag^+ and NO_3^- ions are ----- and -----, respectively.

FURTHER READING

[1] Rakshit, P. C. "*Physical Chemistry*" 7th Edition, Sarat Book House, Calcutta, India, **2014**.
[2] Glasstone, S. "*Textbook of Physical Chemistry*", 2nd Edition, Macmillan, **1958,** Digitized **2009.**
[3] Palit S. R.; De, S. K. "*Practical Physical Chemistry*" Science Book Agency, Kolkata, **1974**.
[4] Daniels, von F.; Mathews, J. H.; Williams, J.W.; Bender, P.; Alberty, R. A.; "*Experimental Physical Chemistry*" McGraw-Hill Book Company INC., New York-Toronto-London **1956**
[5] Athawale, V. D.; Mathur, P. "*Experimental Physical Chemistry*" New Age International (P) Limited Publishers, New Delhi, **2011**.

CHAPTER 9

Some Advanced Physical Chemistry Experiments

STUDY OF CORROSION KINETICS OF IRON IN ACID MEDIUM (CONC. HCl)

Chemicals and Apparatus Required

Iron bar, Conc. HCl, four beakers.

Theory

Corrosive environments have received a considerable amount of attention because of their destruction of materials. One of these environments is the acid solutions which are often used in industry for washing, descaling, and pilling of steel structures, processes which are generally accompanied by considerable digestion of the metal. The information about corrosion rate and kinetic parameters may be helpful in the counted. Activation parameters for some systems can be estimated either from the Arrhenius equation or from the transition state theory.

Chemical kinetics study deals with the rates of chemical processes. Chemical kinetics include an investigation of how different experimental conditions can influence the rate of a chemical reaction and yield information about the reaction mechanism as well as the construction of mathematical models that can describe the characteristics of a chemical reaction.

Iron is widely used in many industries. During industrial processes, such as pickling, etching, acid cleaning, acid descaling, iron is often made to meet aggressive solutions such as acidic, basic solutions. Hence metal is prone to corrosion attacks. The magnitude of corrosion of a metal depends on the concentration of the acidic and basic medium, operating temperature and period of contact, *etc.* various works have the interaction of this medium on the surface of the iron.

Charu Arora and Sumantra Bhattacharya

Procedure

1) An iron bar was purchased from the local market for weight loss measurement. The specimen surface was polished with sandpaper and then dipped in an acid solution of different concentrations.

2) 1 (N), 2 (N), and 3 (N) solutions of hydrochloric acid solutions were prepared from the concentrated HCl solution.

3) The iron bar of known mass was immersed in the prepared solution. The iron bar was immersed in different concentrations of one kind of acid solution for an equal time interval. It was then recovered from the test solution and washed with distilled water, and dried in a hot oven. Finally, the mass of the bar was measured very accurately (Table **9.1**). The difference in weight for each variation was recorded, and from that, corrosion rates and specific reaction rates were calculated.

Observation Table

Experimental Temperature = ...°C.

Table 9.1. Effect of hydrochloric acid concentration.

HCl Concentration (N)	Initial Weight (g)	Final Weight (g)	Period (min)	Weight loss (g)
1 (N)			60	
2 (N)			60	
3 (N)			60	

Kinetics Study

The initial weight of iron specimen and change in weight of specimen at the various instant of time in hydrochloric acid were calculated using first-order rate expression:

$$k = 2.303 \log\left(\frac{\text{Initial weight of specimen}}{\text{Weight of specimen at time t}}\right) \qquad \textbf{(9.1)}$$

The half-life time value was calculated by using the formula $t_{1/2} = \frac{0.693}{k}$. It was observed that the concentration of the hydrochloric acid solution was from 1 (N) to 3 (N). The values of the specific rate constant and the half-life are given in the following Table **9.2**. Based on the results, we may say that the corrosion of iron in an acidic medium follows first-order kinetics.

Observation Tables

Table 9.2. Kinetics of corrosion reaction.

Conc. of HCl (N)	Initial Weight w_1 (g)	Final Weight W_2 (g)	Weight Loss $(w_1 - w_2)$ g	Time Dipped in soln. (min)	Specific Rate Constant (c/s)	Half-Life (min)
1 (N)				60		
2 (N)				60		
3 (N)				60		

Conclusion

From the corrosion study of the iron specimen in 1 (N), 2 (N), and 3 (N) HCl acid solution, it is found that as the concentration of the HNO_3 increases, the weight loss of the iron specimen also increases vigorously. As a result, the specific rate constant also increases rapidly with the increasing concentration of the HCl acid, and the half-life period of the specimen of iron decreases abruptly with an increase in the concentration of the acid.

II. PREPARATION AND CHARACTERIZATION OF SILVER NANOPARTICLES

Chemicals and Apparatus Required

Silver Nitrate, Sodium citrate (capping agent), Sodium borohydride (0.025M), Volumetric flask (100 ml), Beaker (100 ml, 250 ml), Graduated pipette, Burette, Weighing balance, Spatula, UV- vis spectrophotometer.

Theory

The formation of Ag nanoparticles can be observed by a change in colour since small nanoparticles of silver are yellow. Sodium borohydride is used as a reducing agent, and a layer of absorbed borohydride anions on the surface of nanoparticles keeps the nanoparticles separated. The scattering of light caused by spherically shaped colloidal silver nanoparticles shows golden yellow colour. Nanomaterials deal with a process that takes place on the nanometer scale, *i.e.,* from approximately 1 to 100 nm of bulk material made from the same element(s). For example, silver metal is greyish, but colloidal silver synthesized by this method is yellow. The striking effect of nanoparticles on colour has been known since prehistoric times when tiny metal particles were used to colour glass in church windows.

Synthesis of Colloidal Ag

Colloidal silver is made by adding an excess of the reducing agent $NaBH_4$, $NaBH_4$ to $AgNO_3$. Particle size can be determined [1].

$$AgNO_3 + NaBH_4 \text{ ---------------- } Ag + \tfrac{1}{2} H_2 + \tfrac{1}{2} B_2H_6 + NaNO_3$$

Characterization of Colloidal Ag

The sample has to be characterized by UV visible spectroscopy. The presence of metal in the solution is related to a broad absorbance peak at 410nm. The height of the peak gives information about the metallic compound concentration in the medium. In general, as the particles get larger, the absorption maxima shift to a longer wavelength. During the interaction process between electromagnetic radiation and the sample, certain characteristic wavelengths are absorbed by the sample; hence, the intensity of the transmitted light is decreased. The measurement of the decrease in intensity of the radiation is the basis of spectroscopy.

Procedure

1) 0.005 M of $AgNO_3$ solution was made by dissolving 0.085 g of $AgNO_3$ in 100 ml of H_2O in a volumetric flask. From this solution, 100 ml of 0.00025 M $AgNO_3$ was prepared.

2) 100 ml of 0.000125M sodium citrate solution was made by dissolving 0.0032g of sodium citrate in a conical flask/beaker.

3) Dissolve 0.0047 g of $NaBH_4$ in 100 ml of water to prepare 0.00125M sodium borohydrate solution.

4) 10 ml of Ag NO_3 solution and 10ml sodium citrate were mixed, and $NaBH_4$ was added in a drop-wise manner until the colour of the solution changed to yellow.

5) Spectra of the coloured solution were studied using UV-Visible spectroscopy, and the λ_{max} was recorded.

Observation

Stock solution of $AgNO_3$ = ……. M.
Colour of the nanoparticle in the solution = ……..
λ_{max} observed for this solution is = …. nm.

Results

The formation of a pale yellow coloured solution having $\lambda_{max} \approx 420$ nm confirms the formation of Ag nanoparticles.

III. STUDY OF THE EFFECT OF UREA ON PROTEIN STRUCTURE BY UV-VISIBLE SPECTROSCOPY[a]

Requirements

Bovine Serum Albumin (BSA) stock: 7.5 mg/ml, 8 M stock of Urea, Double Distilled Water, Measuring cylinder, Glass Pipette, Pipette, Micropipette.

Theory

Small organic molecules in an aqueous solution can have profound effects on protein stability, structure, and function. The use of these solutions to stabilize or destabilize proteins, depending on the cosolvent, is a common practice. Protein studies are conducted almost exclusively in complex solutions. Chemical denaturation with an agent such as urea is one of the primary ways to assess protein stability, the effects of mutations on stability, and protein unfolding. Despite its widespread use, the molecular basis for urea's ability to denature proteins remains unknown. Urea may exert its effect directly by binding to the protein or indirectly by altering the solvent environment. Most versions of the direct interaction model posit that urea binds to and stabilizes the denatured state (D), thereby corroborate unfolding, but this interpretation does not explain how the protein surmounts the kinetic barrier to unfolding. In this regard, urea could bind with the protein and compete with native interactions, thereby actively participating in the unfolding process. Alternatively, it has been proposed that urea acts indirectly by altering the solvent environment, thereby mitigating the hydrophobic effect and facilitating the exposure of residues in the hydrophobic core. It is also possible that the mechanism of urea-promoted unfolding depends on the urea concentration. The extent of denaturation of BSA can be monitored by measuring the absorbance at 280 nm.

Procedure

1) From the provided BSA stock of 7.5 mg/ml, 500 µl of 0.3 mg/ml BSA was prepared as the working sample.

2) From the provided 8 M Urea stock, the following dilutions were prepared with the protein sample (Table **9.3**).

Table 9.3: Preparation of different dilutions of protein sample.

S. No.	Concentration of BSA (mg/ml)	Volume of Double Distilled H_2O (µl)	Volume of Urea (µll)	Total Volume(µ ll)
Control	0.3	480	0	500
1M	0.3	417.5	62.5	500
2M	0.3	355	125	500
3M	0.3	292.5	187.5	500
4M	0.3	230	250	500
6M	0.3	105	375	500
8M	0.3	0	480	500

Observation

In the 8M urea solution, the peak of the native BSA shifted from 278.4 nm to 276.7 nm, indicating a blue shift.

IV. DETERMINATION OF MAGNETIC SUSCEPTIBILITY OF A PARAMAGNETIC MATERIAL BY USING QUINCKE'S METHOD[b]

Chemicals and Apparatus Required

$MnSO_4.H_2O$, Quincke's tube with a stand, Electromagnet.

Theory

Quincke's method is used to determine the magnetic susceptibility of a diamagnetic or paramagnetic substance in the form of a liquid or an aqueous solution. When an object is placed in a magnetic field, a magnetic moment is induced in it.

The magnetic moment can be measured either by force method, which involves the measurement of the force exerted on the sample by a homogenous magnetic field, or induced method, where the voltage induced in an electric circuit is measured by varying magnetic moments.

The force on the sample is negative of the gradient of the change in energy density when the sample is placed.

$$f = \frac{d}{dx}\left[\frac{1}{2}\mu_o\,(\mu_r - \mu_{ra})\,H^2\right] = \frac{1}{2}\,\mu_o\,(\chi_0 - \chi_a)\,\frac{d}{dx}H^2 \quad \textbf{(9.2)}$$

μ_o= permeability of free space.

The force acting on an element of area A and length d_x of the liquid column is f Ad_x, so the total forces F of the liquid:

$$F = \int f d_x = \frac{A\mu_o}{2}\,(\chi - \chi_a)\,(H^2 - H_0^2) \quad \textbf{(9.3)}$$

Now, F = Ah $(\rho - \rho_a)$ g

$$\text{Or,}\quad \chi = \chi_a + \frac{2}{\mu_o}g\,(\rho - \rho_o)\,\frac{h}{(H^2 - H_0^2)} \quad \textbf{(9.4)}$$

In actual practice χ_a, the density of air ρ_a and H_o are negligible and can be ignored, and the above expression simplifies to,

$$\chi = \frac{2\rho gh}{\mu 0 H2} \quad \textbf{(9.5)}$$

In C.G.S unit, eq (9.4) and (9.5), are:

$$\chi_A = \chi_a + 2g\,(\rho\text{-}\rho_a)\frac{h}{(H^2 - H_0^2)} \quad \textbf{(9.6)}$$

$$\chi = \frac{2\rho gh}{H^2} \quad \textbf{(9.7)}$$

Where ρ, g, h, and H are measured in g/cm^3, cm/s^2, cm, and gauss, respectively.

Procedure

1) The density of the specimen is measured.

2) The Quincke's tube between the poles pieces of the magnet is kept.

3) The liquid in the tube is filled, and the meniscus is set centrally with a pole piece, as shown in the figure.

4) The magnetic field is applied H, and its value is noted from the calibration.

5) The displacement is measured "h" as a function of applied H by h, as a function of H^2 (Table **9.4**).

Observation

Table 9.4. Displacement (h) and applied magnetic field H.

I	H(Gauss)	Reading (mm)	h(mm)	H^2 (Gauss)^2

Calculation

From the graph,

Slope, $\frac{h}{H^2}$ = ------

$\chi = \frac{2\rho gh}{H^2}$ = -------

V. TO STUDY THE HALL EFFECT[b]

Chemicals and Apparatus Required

Hall instrument set up, Germanium rod.

Theory

According to the Hall Effect principle, when a conductor or semiconductor with current flowing in one direction is introduced perpendicular to the magnetic field a voltage could be measured at right angles to the current path. The effect of getting a measurable voltage, as told above is known as the Hall Effect. To properly impact items such as pipes or tubes, Hall Effect probes work with magnetic Fluor lockage to properly impact items such as pipes or tubes.

This is a way of testing such items and being able to spot potential corrosion, corrosion, or pitting. This is specially used in steel items and can give important information about life or safety. Since we know that the Hall Effect is used to measure magnetic fields and power overall. To measure these magnetic fields, we need a probe or sensor to do the hard work. Because the probe only gives off a low level of signals and it needs to be pushed further to be able to read it. The hall voltage that develops across a conductor is directly proportional to the current, to the magnetic, and to the nature of the particular conducting material itself and is inversely proportional to the thickness of the material in the direction of the magnetic field.

Procedure

1) Connected the constant current source to the solenoids.

2) Four probes are connected to the gauss meter and placed in the middle of the solenoids.

3) Switch on the gauss meter and constant current source.

4) Vary the current through the solenoid from 1 Amp for set-1 and 2 Amp for set-2.

5) Switch off the gauss meter and constant current source.

6) Fix the hall probe as a wooden stand. Connect green wires, constant current generator (CCG), and red wires to the milli-voltmeter in the Hall Effect apparatus.

7) The hall probe is then placed at the middle of the two solenoids replacing the fixed four-probe or magnet.

8) Switch on the constant current source.

9) Increase gradually the current (I) from CCG at around 0.5 A & measured the corresponding hall voltage (V_Z) for two acts.

10) Then calculated the hall-coefficient (R_H) from the equation

$R_H = V_H t/IB$ t = thickness of the sample, B = magnetic field

Mobility, $\mu = R_H \sigma$, Current density $H = 1/R_H e$ $[R_H = 1/H_e]$

11) Plot the graph current in milliampere (along X-axis) Vs. The magnetic field in Gauss (along Y-axis) and Magnetic field (Gauss) Vs. Hall voltage (V).

12) Both will be straight-line graph passing through the origin and from the slope of the graph calculate V_h/H.

VI. SYNTHESIS OF A NYLON-6,6 POLYMER

Introduction

Nylon-6,6, a synthetic polymer, was synthesized and discovered to have commercial properties in the 1930s by Wallace Caruthers and co-workers at a DuPont research laboratory. The same industry began the industrial-scale production of nylon and their first commercial nylon product, nylon stockings, was first unveiled in the market in 1940. Since then, various other forms of nylon, like nylon-6,10 have been developed. Nylon is presently found in numerous commercial products like clothing, parachutes, ropes, toothbrush bristles, *etc.* Additionally, many machined parts, such as the frame of the Glock handgun are made of a nylon composite. Polymers are large (high molecular weight) molecules made by repetitively bonding together many smaller units called monomers as shown in

Fig. (**9.1a**). To represent the many units that are present, a single monomer unit is placed in parenthesis with a subscript n to represent some unknown number of units. One of the most commonly encountered polymers is polystyrene (Styrofoam), which is made up of many repeating styrene units as shown in Fig. (**9.1b**). In the case of styrene, the monomer units are each joined by new single bonds (in red). Because the styrene double bond took part in the polymerization reaction process, the styrene subunits within the polymer no longer possess the alkene functionality.

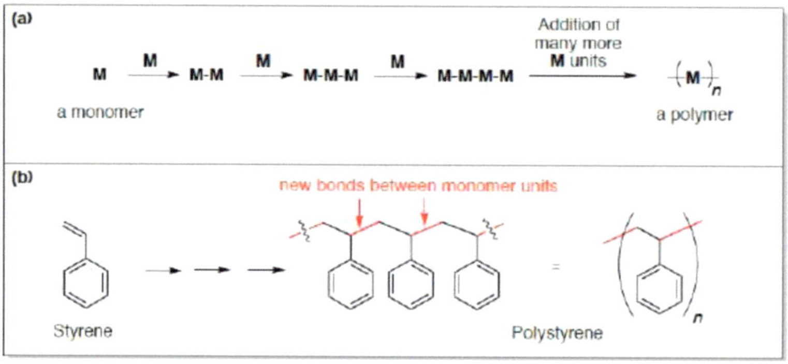

Fig. (9.1). Reaction scheme for polymer formation. **(a)** General scheme for polymer formation, **(b)** Scheme for polystyrene formation

Polystyrene is an example of a chain-growth polymer because it is synthesized by a chain reaction where some initiator adds to the first molecule of monomer to yield a reactive species which in turn reacts with a second molecule of monomer and so on (Fig. **9.2**). Also, polystyrene is classified as a homopolymer because it consists of identical repeating units.

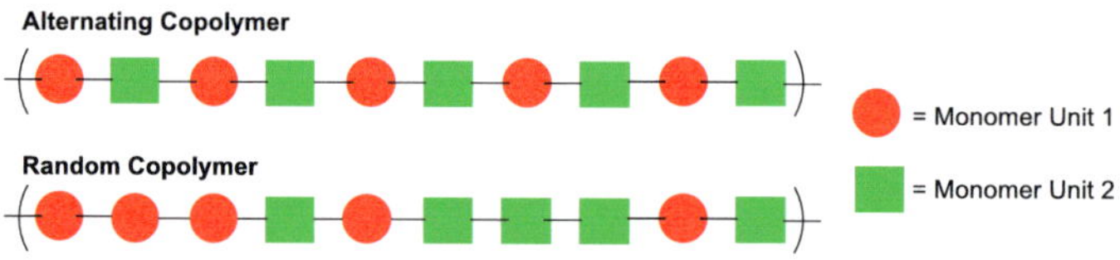

Fig. (9.2). Schematic description of Alternating and Random polymers.

Nylon 6,6 is an example of an alternating co-polymer because it contains alternating diamine and dicarbonyl units (Fig. **9.3**). The 6,6 numbering refers to the number of carbon atoms in the diamine and dicarbonyl units, respectively. Another frequently encountered synthetic nylon polymer is Nylon 6,10. This polymer also has six-carbon diamine units, but the dicarbonyl unit contains ten carbon atoms.

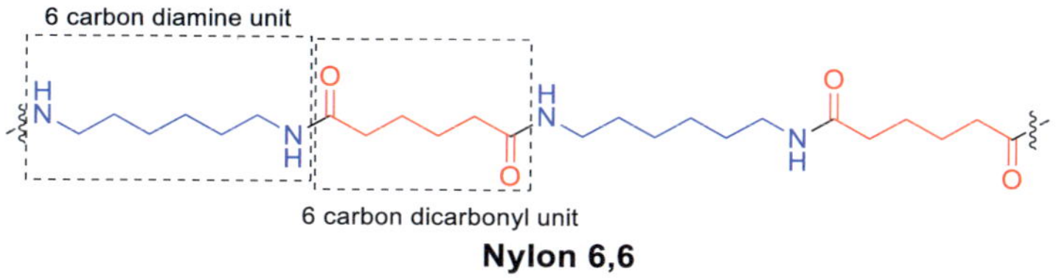

Fig. (9.3). Formation of Nylon 6,6 polymer.

Chemicals Required

Adipic acid, Hexamethylene diamine, Sodium hydroxide, Alcohol.

Procedure

1) Place 10 ml of a 5% solution of adipoyl chloride in cyclohexane in a 50-ml beaker. Place 10ml of a 5% aqueous solution of hexamethylenediamine in a separate 150-ml beaker. Add 7 drops of 20% sodium hydroxide solution to the hexamethylenediamine solution. Slightly tilt the beaker containing the aqueous diamine solution and carefully pour the adipoyl chloride solution down the wall of the beaker to form two layers. If the solutions are poured together too vigorously, the mixing of the two solutions will occur and it will be impossible to form a nice film at the interface. A polymer film should form immediately at the liquid-liquid interface. Using forceps gently pull a solid mass of polymer from the center of the liquid-liquid interface. At this point, you should be able to continue pulling polymer from the flask to produce a long rope of nylon. You can wind the rope around a test tube as it is being continuously pulled from the flask.

2) Alternatively, you can slowly walk along the bench as you continuously pull a long strand of rope from the flask. You should be able to get a rope that is several feet long. Rinse the rope several times with water and lay it on a paper towel to dry.

3) Once you can no longer pull nylon rope from the flask, vigorously stir the remainder of the two-phase solution to see if any additional polymer forms. Decant the liquid away from any polymer remaining in the flask. Wash this polymer first with 50% aqueous ethanol and then with water and allow it to dry.

Observations

- The total mass of polymer formed. (The polymer must be thoroughly dried).
- The mass of your longest piece of nylon rope.
- The length of your longest continuous piece of nylon rope.
- Describe the color, appearance, texture, and shape of the nylon rope.
- Stretch test for tensile strength. Cut a 1-inch section of your nylon rope. Stretch it along a ruler to the point where it breaks. Record the maximum length that the nylon stretched to before breaking.

ACKNOWLEDGEMENTS

[a]The practical is designed by Dr. Amlan Das, Project Manager. National Institute of Biomedical Genomics., Kàlyani, West Bengal, India. PIN-741251.

[b]This Experiment is designed by Mr. Happy Mondal, Department of Physics, NIT Sikkim, Barfung Block, Ravangla, South Sikkim. India, PIN-737139.

FURTHER READING

[1] Das, R; Jagannathan, R; Sharan, C.; Kumar, U.; Poddar, P; *J. Phys. Chem. C* **2009**, 113, 21493.
[2] Cooper, A. "*Biophysical Chemistry, Royal Society of Chemistry*", **2004**.
[3] Rakshit, P. C. "*Physical Chemistry*" 7th Edition, Sarat Book House, Calcutta, India, **2014**.
[4] Levine, I.N. "*Physical Chemistry*" 6th Edition, Tata Mc. Graw Hills, New Delhi, India, **2011**
[5] Atkins, P.W.; de Paula, J. "*Atkins' Physical Chemistry*" 11th Edition, Oxford University Press, **2018**.
[6] Kittel, C. "*Introduction to Solid State Physics*" 8th Edition, John Willey & Sons Inc. **2005**.
[7] Glasstone, S. "*Textbook of Physical Chemistry*", 2nd Edition, Macmillan, **1958,** Digitized **2009.**
[8] Palit S.R.; De, S.K. "*Practical Physical Chemistry*" Science Book Agency, Kolkata, **1974**.
[9] Daniels, von F.; Mathews, J. H.; Williams, J.W.; Bender, P.; Alberty, R. A.; "*Experimental Physical Chemistry*" McGraw-Hill Book Company INC., New York-Toronto-London **1956**.
[10] Athawale, V. D.; Mathur, P. "*Experimental Physical Chemistry*" New Age International (P) Limited Publishers, New Delhi, **2011**.

CHAPTER 10

General Introduction of Making Structure of Molecules using Computer

INTRODUCTION

In this chapter, we shall execute some experiments without consuming any chemicals, we will be using a computer and necessary software (S/W) instead. We can calculate various properties of atoms, molecules, polymer, periodic systems, *etc.* Although the calculation of properties of molecules is discussed here, computational studies of the rest of the mentioned systems are beyond the scope of the book.

In the last two decades, computational chemistry has become a very demanding tool for experimental chemists to corroborate several chemical phenomena at the molecular level. Using computational chemistry, one can predict several properties of chemical systems, like *in silico* (Since computer chips consist of silicon semiconductors, so computational jobs are often known as *in silico* work) prediction of spectroscopic properties, response properties (optimized structure, electrical properties, magnetic properties), excited-state properties, adsorption, *etc.*

Here, we shall present a brief introduction to drawing the structure of small molecules using standard software and some preliminary quantum chemical calculations of some small molecules. We have also discussed how to study the geometry of a crystal from a Crystallographic Information File (CIF).

Nowadays, several software packages are available to compute energy accurately and various properties of molecules that involve energy. Among several available software packages to design molecular structures, ChemDraw [1], GaussView [2], Molden [3] are frequently used. Among these software packages, the first two are paid software (One has to purchase), while the last one is free to use. Similarly, to calculate energy and response properties of molecules using quantum chemical methods, Gaussian 09 [4], TURBOMOLE [5], GAMESS [6], ORCA [7], *etc.*, are widely used. Among these software packages Gaussian, TURBOMOLE are not free, but the GAMESS and ORCA, *etc.*, software suits are free to use for academic purposes. It is necessary to mention that all the software packages are not available in Windows Operating Systems (OS). To the best of our knowledge, Gaussian,

Charu Arora and Sumantra Bhattacharya

GAMESS, and TURBOMOLE are available in both Windows and UNIX OS, but the rest are available in UNIX OS.

To calculate the energy and properties of molecules using quantum chemical methods, the Hartree-Fock (HF) [8] method is widely used. However, the Hartree-Fock method can give exact energy up to 95% accuracy [8] because the correlation energy is not taken into account properly. To get very accurate results, one has to adopt several correlated methods, like Configuration Interaction (CI) [8], Many-Body Perturbation Theory (MBPT) [8], coupled cluster (CC) [9], Density Functional Theory (DFT) [10], *etc.*

In this chapter, we shall discuss a brief outline of the HF method and the basis set. Using ChemDraw and GaussView how one can draw structures of molecules, and finally, we shall show how to calculate the energy of small molecules using the Gaussian software package (g09).

Hartree-Fock Method and Basis Set

To evaluate the energy of an atom or a molecule using quantum mechanics, solving the equivalent time-independent Schrödinger equation is required to determine wave-function (Ψ) of the system, which is:

$$\widehat{H}|\Psi\rangle=\varepsilon|\Psi\rangle \tag{10.1}$$

$\widehat{H}$ is the Hamiltonian or total energy operator. $\widehat{H}$ consists of the kinetic energy of constituent particles and potential energy resulted out from various interactions among them. The potential energy can be of attractive interaction, which arises because of interaction between oppositely charged particles, like attractive interaction between electron and nucleus or repulsive interaction which arises because of interaction between same charged particles, *e.g.*, electron-electron repulsion, nucleus-nucleus repulsion. Notably, the operators are represented by the symbol '^', but to simplify notation, we drop the symbol.

To represent the Hamiltonian (total energy) operator for a molecule, let us assume a molecule consists of M number of protons and N number of electrons. The Hamiltonian can be expressed as:

$$H=-\frac{1}{2}\sum_{A=1}^{M}\frac{1}{M_A}\nabla_A^2-\frac{1}{2}\sum_{i=1}^{N}\nabla_i^2-\sum_{i=1}^{N}\sum_{A=1}^{M}\frac{Z_A}{\|r_i-R_A\|}+\sum_{\substack{i,j\\i<j}}^{N}\frac{1}{\|r_i-r_j\|}+\sum_{\substack{A,B\\A<B}}^{N}\frac{Z_AZ_B}{\|R_A-R_B\|} \tag{10.2}$$

The successive first two terms of the right-hand side of Equation (10.2) are the kinetic energy of the nuclei and electron, respectively. R_A and r_i are the spatial coordinates of nuclei A and electron *i*, respectively. The third term is electrostatic energy of attraction between electron i and nucleus A, Z_A is the charge of nucleus A. Among the five terms, the first three terms represent attractive force. On the other hand, the last two terms are the repulsive force. Between the last two terms, the first one is repulsive electrostatic force between electron *i* and *j*, the restriction i<j is imposed to avoid double counting; another option is to multiply by a factor 1/2. The same logic is imposed for the last term also, which arises because of electrostatic repulsion between nucleus A and B of charge Z_A and Z_B. Ψ is the wave function of the system and is a function of space and spin coordinates of nuclei, denoting the combined space-spin coordinate of ith electron by $x_i = (r_i, \xi_i)$. Where, ξ_i, is the spin of ith electron. The N electron-M nuclear system wave function (Ψ) can be represented as:

$$\Psi(R_A, ..., R_M, x_1, ..., x_N)$$

Solution of the eigenvalue Equation (10.1) gives stationary state energies (When the first derivative of energy with respect to time is zero, the state is considered as a stationary state) and wave-functions.

Solution of Equation (10.1) is exactly solvable only for the hydrogen atom, but it is impossible for many-electron systems because of the repulsion terms. Hence, to solve Equation (10.1) for many-electron systems, we should adopt approximation techniques.

The nucleus is almost 1836 times heavier compared to electrons, so the velocity of the nuclei should be much lower than that of electrons. Thus, the motion of the nuclei can be frozen when we are interested in studying the electronic structure. This approximation is popularly known as the Born-Oppenheimer (BO) approximation [11]. Thus, by using BO approximation, we can drop the first and last terms of Equation (10.2). The rest term of the total Hamiltonian can be considered as electronic Hamiltonian (H_{el}), and can be expressed as:

$$H= -\frac{1}{2}\nabla_i^2 - \sum_{A=1}^{M} \frac{Z_A}{\|r_i - R_A\|} + \sum_{\substack{i,j \\ i<j}}^{N} \frac{1}{\|r_i - r_j\|} \quad \textbf{(10.3A)}$$

$$H_{el} = \sum_{i=1}^{N} h(r_i) + \sum_{i<j}^{N} g(\|r_i - r_j\|) \quad \textbf{(10.3B)}$$

Where $h(r_i)$ and $g(\|r_i - r_j\|)$ are represented as:

$$h(r_i) = -\frac{1}{2}\nabla_i^2 - \sum_{A=1}^{M} \frac{Z_A}{\|r_i - R_A\|} \quad \textbf{(10.4A)}$$

$$g(\|r_i - r_j\|) = \frac{1}{\|r_i - r_j\|} \quad \textbf{(10.4B)}$$

In Equation (10.3B), the first term on the right-hand side is known as one body operator and represents the coordinate of electron i, $h(r_i)$ is known as the core operator and represents the Hamiltonian of an individual electron at the given nuclear configuration, no other interaction term is considered. The second term is known as electrostatic repulsion force between electron i and electron j. It is a two-body operator. Solution of Equation (10.1) at a fixed nuclear geometry, with the above Hamiltonian, used in Equation (10.3) gives wave functions $\Psi_{el}(x_1,....,x_N;R_1,....,R_M)$ for different electronic states with corresponding electronic energies. $E_{el}(R_1,....,R_M)$. The addition of nuclear terms to electronic energy provides a potential energy surface (PES).

The term x_i, depends not only on the spatial coordinate, it is also dependent on the spin part. According to Pauli's exclusion principle, no two electrons can have the same quantum number. Thus, in a spatial orbital maximum of two electrons can be accommodated. Since the electrons are Fermions, that's why electrons are antisymmetric with respect to the interchange of space-spin coordinates of any two-electron. Mathematically anti-symmetry can be represented as:

$$\Psi(x_1,x_2,....,x_N) = -\Psi(x_2,x_1,....,x_N) \quad \textbf{(10.5)}$$

Instead of the above representation, we can represent the same in determinant form, which is popularly known as Slater determinant and can be written as:

$$\Psi(x_1,x_2,...,x_N) = \frac{1}{\sqrt{N!}} \begin{vmatrix} \chi_1(x_1)\chi_1(x_2).......\chi_1(x_N) \\ \chi_2(x_1)\chi_2(x_2).......\chi_2(x_N) \\ \cdots\cdots\cdots\cdots\cdots \\ \cdots\cdots\cdots\cdots\cdots \\ \cdots\cdots\cdots\cdots\cdots \\ \chi_N(x_1)\chi_N(x_2)....\chi_N(x_N) \end{vmatrix} \quad \textbf{(10.6)}$$

In shorthand notation, we can express a normalized Slater determinant as:

$$\Psi_I(x_1,x_2,....,x_N) = \frac{1}{\sqrt{N!}} \left|\chi_i(x_1),\chi_j(x_2),....,\chi_k(x_N)\right\rangle \quad \textbf{(10.7)}$$

Here the normalization constant is included and shows the diagonal elements of the determinant.

In the BO approximation, we have shown that in Equation (10.3A), an inter electronic repulsion term is present, making the solution impossible to solve, as the term $\left(\|r_i\text{-}r_j\|\right)^{-1}$ can never be represented as: $(r_i)^{-1} - (r_j)^{-1}$. Hence, it is impossible to represent the Hamiltonian as $H = H_i + H_j$. Thus, to handle this term, we have to adopt further approximations. In this context, the Hartree-Fock (HF) approximation is the most popular method.

The HF approximation is a milestone in the electronic structure theory to describe an approximate solution of the electronic part of the Schrödinger equation. In the HF method the $1/r_{ij}$ the term is handled very brilliantly.

The basis of the HF theory is that the stationary states of many electronic systems, particularly in the ground state of a closed shell (All the electrons are paired) atoms and molecules can be described by single Slater determinant Equation (10.6):

$$\Psi_I(\vec{x}_1,\vec{x}_2,....,\vec{x}_N) = \frac{1}{\sqrt{N!}}\left|\chi_i(\vec{x}_1),\chi_j(\vec{x}_2),....,\chi_k(\vec{x}_N)\right\rangle$$

In this approximation, each electron is assumed to be independent of each other, *i.e.*, electrons are assumed to move in a spherically averaged inter electronic repulsion potential. This approximation takes care of $1/r_{ij}$ part. The HF method is known as the independent particle model. The spin orbitals are approximately varied by constraining them only to the extent that they remain orthonormal, *i.e.*, $\langle\chi_a|\chi_b\rangle=\delta_{ab}$ until the minimum energy is achieved. In this way, we get the best spin orbitals, which minimize the energy. This procedure leads to an integrodifferential equation, which is known as Hartree-Fock equations. In the HF equation, the Fock matrix (f(x)) consists of two parts:

(1) One electron operator. It includes the kinetic energy of electrons (T_e) and potential energy of electron with nucleus (V_{ne}).

(2) Two-electron operator ($V_{HF}(x)$). The two-electron operator takes care of the most important $1/r_{ij}$ term. The Hartree-Fock equation can be written as follows:

$$f(x)\chi_a(x)=\varepsilon_a\chi_a(x) \quad \textbf{(10.8)}$$

$$f(x)=T_e+V_{ne}+V_{HF}(x)$$

$$=-\frac{1}{2}\nabla^2 - \sum_{A=1}^{M} \frac{Z_A}{\|r-R_A\|} + V_{HF} \quad \textbf{(10.9)}$$

$$V_{HF}(x) = \sum_{j=1}^{N} J_j(x) + \sum_{j=1}^{N} K_j(x) \quad \textbf{(10.10)}$$

$$J_j(x)\chi_i(x) = \int dx' \frac{\chi_j^*(x')\chi_j(x')}{|x-x'|}\chi_i(x) \quad \textbf{(10.11)}$$

$$K_j(x)\chi_i(x) = \int dx' \frac{\chi_j^*(x')\chi_i(x')}{|x-x'|}\chi_j(x) \quad \textbf{(10.12)}$$

f(x) is the Fock operator. $V_{HF}(x)$ is the Hartree-Fock potential. $V_{HF}(x)$ consists of two parts: (1) The Coulomb term and (2) The exchange term.

The Coulomb term is expressed by $J_j(x)$. Expression of $J_j(x)$ is given in Equation (10.11). The physical significance of the Coulomb term is that it is a spherically averaged potential experienced by an electron due to the motion of all the other electrons.

The exchange term is denoted as $K_j(x)$. The expression of $K_j(x)$ is given in Equation (10.12). There is no classical mechanics analogue regarding the physical significance of the exchange potential. Since wave function is antisymmetric in nature, so the exchange term is introduced in quantum mechanics.

In the integrodifferential HF method, Equation (10.8) $\chi_a(x)$ exists on both sides. Thus, to solve the equation, an iterative procedure is adopted. To solve this equation, an initial guess of spin orbitals is used to calculate the approximate HF potential. Fock matrix is used to obtain a new set of spin-orbitals. This procedure is repeated until some self-consistency is achieved between successive iteration. That's why this method is known as the self-consistent field (SCF) method.

When Equation (10.8) to Equation (10.12) are solved, it yields an orthonormal set of spin orbitals. (χ_i, i = 1, 2, ..., ∞), with corresponding orbital energy, ε_i. The N spin orbitals (χ_a, a = 1, 2,....,N), with lowest energies are referred to as occupied orbitals (hole), and the remaining set of spin-orbitals (χ_r, r = N+1, ...,∞) are referred to as unoccupied or virtual orbitals (particles).

The HF approximation is in good agreement for the ground state of non-degenerate many-electron systems, and equations are exactly solvable as non-linear integrodifferential equations only for atoms. For molecules, orbitals involved are centred at different nuclei. For this reason, the explicit two-electron interaction term is difficult.

To get rid of the problem, Roothan introduced the concept of expansion of basis set functions, which is a finite set of spatial basis functions $\{\phi_\mu(r),\mu=1,..,k\}$. Basis functions are introduced to expand the spatial part of the spin-orbitals, which converts an integrodifferential form of HF equation into a matrix eigenvalue HF equation for the expansion coefficient. Using the iterative SCF procedure, the equations are solved.

Using the basis set, the HF equation can be written as:

$$FC=SC\varepsilon \quad \textbf{(10.13)}$$

Here **F** is the Fock matrix. **C** is the coefficient matrix. **S** is the overlap matrix, and **ε** represents the orbital energy.

The advantage of HF approximation is that it replaces the complicated many-electron problem with an independent particle picture by treating electron repulsion in a spherically average manner, leading to very popular molecular orbital (MO) theory, where electrons occupy different orbitals. This is the first accurate approximate method to approach modern quantum chemistry.

Details of the Hartree-Fock method are not discussed here. For details, please see Ref. [8].

Basis Set

To solve Equation (10.13), one should use a basis set, which defines a space, through which is a collection of vectors defined over a space of our interest. In quantum chemistry, a basis set refers to the set of non-orthogonal one-particle wave functions to describe the shape of atomic orbitals (AOs), which are further used to construct molecular orbitals (MOs) by using the linear combination of atomic orbitals (LCAO) using the concept that a wave function can be represented as a linear combination of atomic orbitals as $\Psi=\sum_i \mathbf{C_i\Phi_i}$. The AOs are obtained by solving the hydrogen-like equation, which is Slater type functions or Slater type orbitals (STO) ($\mathbf{S_{nlm}}$) of the general form:

$$S_{nlm}(r,\theta,\phi) = \frac{(2\zeta)^{n+1/2}}{[(2n)!]^{1/2}} r^{n-1} e^{-\zeta r} Y_l^m(\theta,\phi) \quad \textbf{(10.14)}$$

Here, n, l, m are three quantum numbers, r is the radial part and θ and φ are angular parts, respectively. ζ is the orbital exponent, controls the width of the orbital, chosen to minimize energy. The STO type functions are very accurate, yet evaluations (computation) of these functions are very expensive, especially when many nuclei are involved, like molecules. To get rid of this problem use of Gaussian type functions instead of Slater orbitals make the computation for the same type of problems becomes easier. General expression for Gaussian function ($G_{nl,}$) or Gaussian type orbital (GTO) is:

$$G_{nlm}(r,\theta,\phi)=N_n r^{n-1} e^{-\alpha r^2} Y_l^m(\theta,\phi) \tag{10.15}$$

α plays the role of controlling the width of the orbital, and thus, it is also necessary to optimize. To minimize the error due to the introduction of Gaussian type orbital, often a hybrid of STO and n number of GTO functions are used to define a basis set. This is known as the STO-nG basis set, *e.g.,* 6-31g basis set, which we use here. Details of the basis set can be found in [8, 12, 13].

In this chapter, we shall use a 6-31 G basis set. Here, 6 indicates the core atomic orbital (inner-shell orbital) of an atom which is represented by a sum of 6 Gaussian types of functions. The hyphen denotes a split-valence basis set, which indicates that 2s and 2p orbitals are represented by a couple of Slater orbitals. The smaller one, *i.e.,* the 2s orbitals, are represented by 3 Gaussian functions, and the larger orbital, *i.e.,* the 2p orbital, is represented by a 1 Gaussian function. Also, note that we can add polarization and diffuse functions to the basis set whenever it is necessary. In conclusion, we say that we can keep on adding functions to the basis set and make the basis set larger, but on the other hand, increasing the basis functions will increase the computational time; hence the proper basis set should be chosen properly.

I. DRAW THE STRUCTURE OF MOLECULES IN COMPUTER

The structure of a molecule can be simplistically drawn in a computer by using the "ChemBioDraw Ultra 12.0" and "Gaussview" software package. Between these two, the first one will give a two-dimensional representation while the three-dimensional structure can be sketched using the Gaussview software package. In this section, we shall give a brief description of these two methods.

I.1. Drawing a Structure in a Computer Using 'ChemBioDraw Ultra 12.0'

Availability of ChemBioDraw Ultra 12.0

The software is available in Windows and Mac operating systems and can be purchased through http://scistore.cambridgesoft.com/chemdraw/. The advantage of this S/W is that the structures can be drawn two-dimensionally but can be visualized in three-dimension also by using ChemBio3D.

Drawing a Structure

When ChemDraw software is launched, the screen appeared as displayed in Fig. (**10.1**):

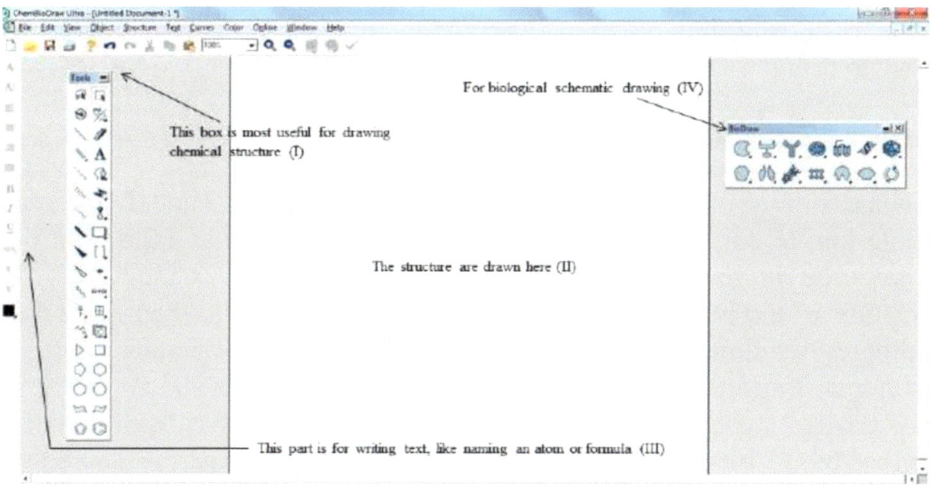

Fig. (10.1). The first screen for ChemDraw.

The marked parts are levelled by roman numerals. Here we shall describe three of them with an illustration of how to draw a structure in ChemDraw. The fourth part will keep untouched, as this is beyond the scope of the book.

(I) Tool: This section is most useful in drawing a chemical structure in two dimensions. The final structure can also be viewed in three-dimension by using the 'chemBio 3D ultra' software package. Looking at the tool section from top to bottom, the first two are used for marking a particular portion. Suppose we want to draw toluene structure and want to mark methyl group, then we first select either of them and mark on our desired portion. In the second row, using the first tool (structure perspective) from the left, we can get the perspective orientation of the drawn molecule. Another part is helpful while doing fragmentation of the system. The third row contains making a single bond and eraser. The usefulness of both is quite clear. In this way, down the column of the toolbar, we can find several utility tools to draw the chemical structure in two-dimension.

(II) The structures are drawn in this part.

(III) Part helps us to make the structure fine-tuned, like making proper size, font, colour, *etc.*

(IV) Section is very useful for biochemists or biologists.

However, we focus our aim only to draw chemical structures so we confine our focus only to that portion. Details discussion on the drawing structure of a molecule using chem draw is beyond the scope. More use of the software will become friendlier to use.

I.2. Drawing Structure of a Molecule Using Gauss View

When gauss view software opens, it appears like Fig. (**10.2**). The left one is used as a building tool to edit the structure of the desired molecule and the right window (blue coloured in the figure and known as active view window) is for the visualization of the built-up molecule. It is to be noted that when more than one blue coloured window is opened, the building tool always operates on the active view window. Various useful tools in GaussView software is shown in Fig. (**10.3**). To draw desired structure, one should choose the proper atom from the atom list, symbolized by ^{6}C, by clicking, the list of atoms will appear and the desired valence will also appear below table as shown in Fig. (**10.4**).

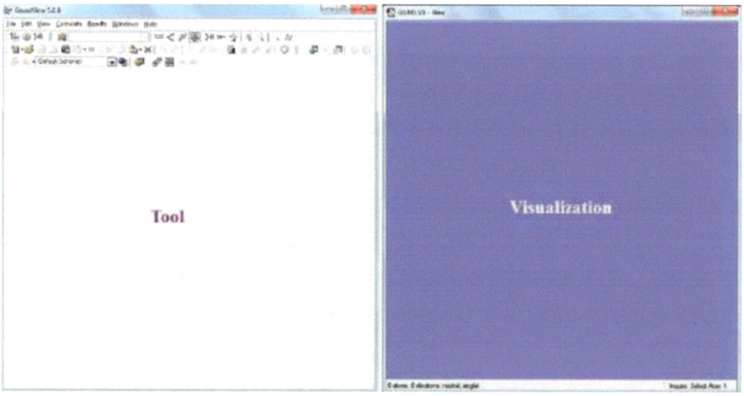

Fig. (10.2). Launching of the Gauss view window.

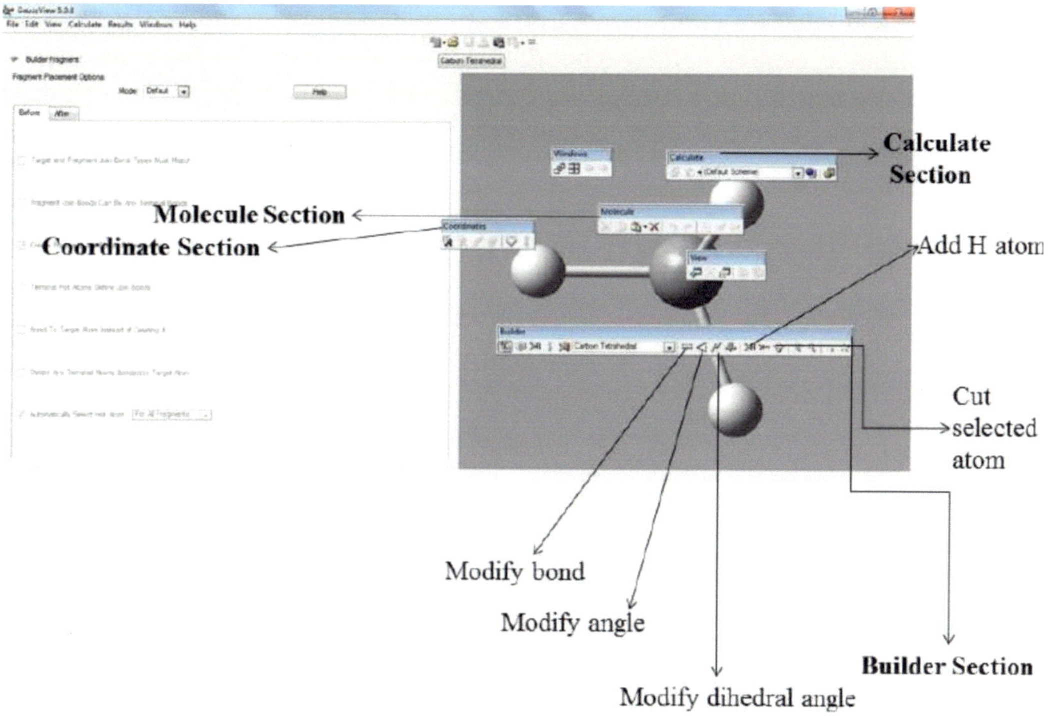

Fig. (10.3). various utility tools in gauss view.

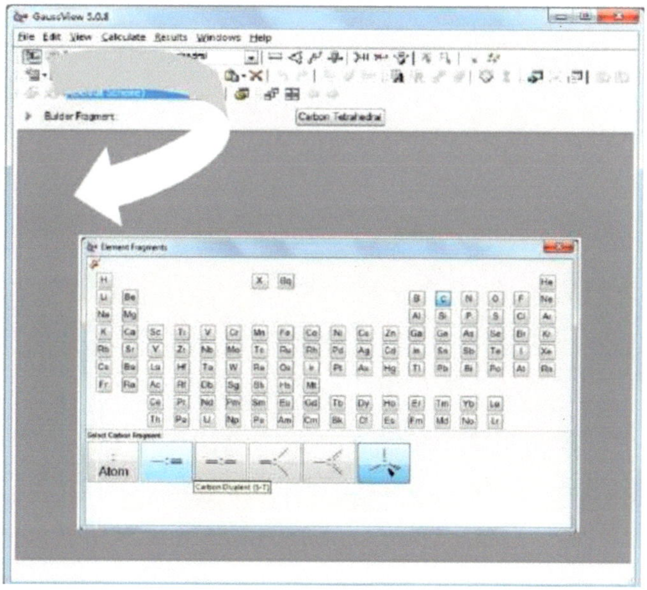

Fig. (10.4). Elements list in Gauss view.

First, choose a single atom and then choose another atom in a similar way and set their bond distance, type of bond (no bond, single, double, triple, or fractional bond) from the builder section. In a similar way, functional groups or fragments or cyclic structures can also be added. List of functional groups and rings displayed in Figs. (**10.5 and 10.6**) below:

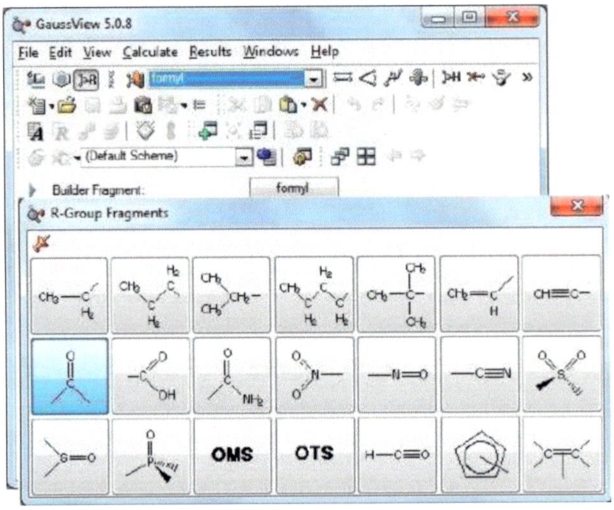

Fig. (10.5). Functional group fragment list in gauss view.

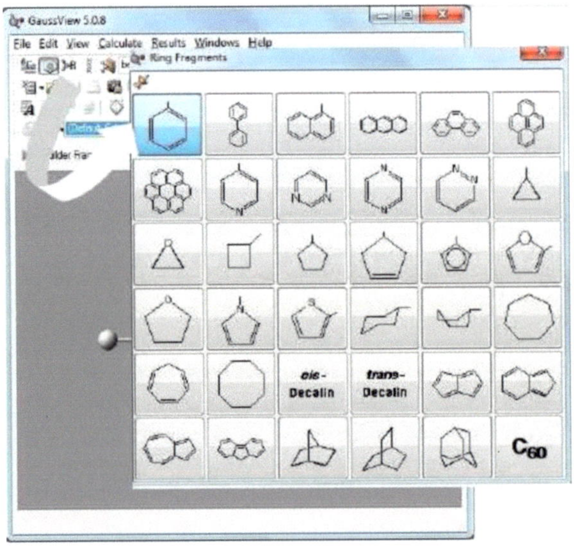

Fig. (10.6). Ring fragment list in gauss view.

To construct a molecule using gauss view software, one should draw the initial structure by choosing the appropriate atom from the atom list as shown in Fig. (**10.4**) and then the change bond length, bond angle, and dihedral angle between atoms manually by clicking the properly as indicated in Fig. (**10.3**).

Making a Gaussian Input File

The Gaussian input file can be made by using both from the graphical user interface (GUI) as well as from the command prompt.

If GUI is used then, one should first draw the desired structure of the molecule, then click on the calculate button in the tool window and select Gaussian calculation setup (ctrl+G). This will give various options. The user has to choose as per requirements. Several options are shown in Fig. (**10.7**).

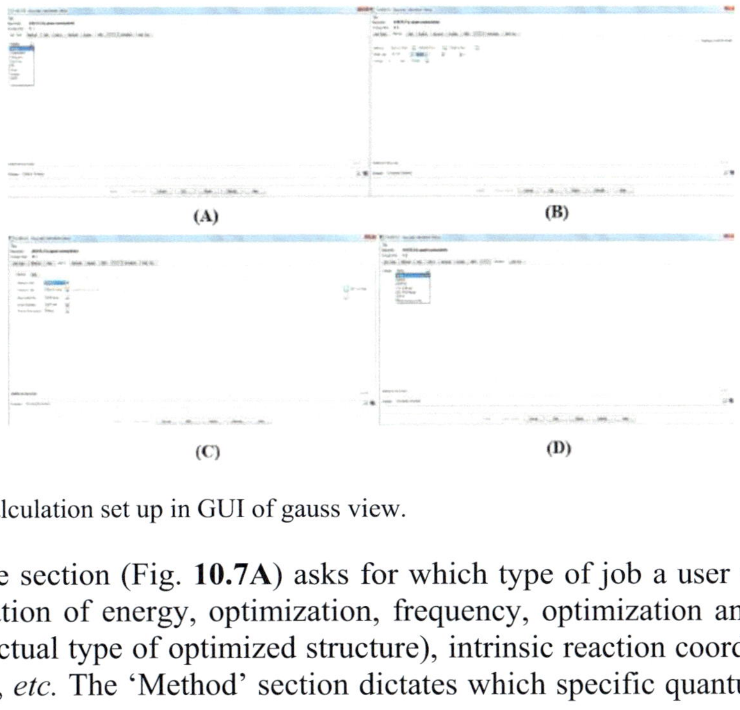

Fig. (10.7). Calculation set up in GUI of gauss view.

The job type section (Fig. **10.7A**) asks for which type of job a user wants to do, like, calculation of energy, optimization, frequency, optimization and frequency (To check actual type of optimized structure), intrinsic reaction coordinate (IRC), Scan, NMR, *etc.* The 'Method' section dictates which specific quantum chemical method to use.

The Method subsection (Fig. **10.7B**) guides which method to use, say the Ground state, ZINDO (semi-empirical method), Time-dependent self-consistent field (TD-SCF) method, Equation of motion coupled cluster (EOM-CC), *etc.* The Ground-state method has several subsections, like, Hartree-Fock, semi-empirical, density functional theory (DFT), coupled cluster truncated at singles doubles excitation (CCSD), *etc.* The third subsection gives information about spin, which is dependent on the overall spin of the molecule and accordingly set will come, like restricted (closed-shell systems, *i.e.,* all electrons are paired up), unrestricted (open-shell systems, where all electrons are not paired up), restricted open. Here a discussion of all these is beyond scope of this book, we confine our discussion only on the restricted Hartree-Fock method. Choice of proper basis sets is an important task for quantum chemists. The second row of the method section provides a list of basis sets, even external basis sets can also be used. Based on the type of calculation, one should choose the basis set properly.

The title part is not so important. The user can write the name for the future track.

The Link section provides information about details of computational aspects like memory limit. This controls the amount of dynamic memory to be used by Gaussian. The default value is 256 MegaBytes (MB), which can be changed by typing the value of memory in the memory limit section with a proper unit of memory (KB, MB, MW, GB, *etc.*, here K, W, G stand for Kilo, words, Giga, respectively. Note 8 bit = 1 byte, 1024 byte = 1 Kilobyte (KB), 1024 KB = 1 MegaByte (MB), 1024 MB = 1 Gigabyte (GB) and so on). One should keep in mind that requesting more memory than the actual available memory of your computer will lead to a slowdown of the calculation. Several parameters Gaussian save in the checkpoint file, which is an unformatted file but is useful for post-processing of various calculations. Thus, saving a checkpoint file is necessary, and the extension is 'chk' (*e.g., filename*.chk). It is advisable to use the default name option in the checkpoint file name for future tracking. If the Gaussian is installed in parallel mode, it is advised to run your Gaussian program in parallel for faster calculations. Gaussian performs best over 4, 8, 12, and 16 processors. So, it is advised if Gaussian is installed in a parallel processor, please put the value of shared processors accordingly (4, 8, 12, or 16).

Since the iterative process is adopted in quantum chemical calculations. Thus, a suitable guess is very important. In Gaussian Harris guess is used, where the Harris function [14] is diagonalized for initial guess. Other than this, the various initial guess is also available in Gaussian software, like Hückel, core, INDO, *etc.* We shall explore the default option only.

To use solvent correction, one can use the solvent model, which is available in the solvent section (Fig. **10.7D**), either of the solvent models. If one has chosen a particular solvent model, a list of solvents (water, methanol, ethanol, THF, DMSO, acetone, *etc.* will appear. As of requirement, a solvent can be chosen.

Once all these are ready, just click on edit and save the structure. It will save as *filename*.com or *filename*.gjf format. This file can be run from the GUI unit or by a command prompt.

The gaussian input file can be made using the command prompt also. To make a Gaussian input file one should open a file by *filename*.com or *filename*.gjf. In the file, please write the following as shown in Fig. (**10.8**) (Here we have chosen the example of water):

%nprocshared=4: Indicate number of processor shared (For single processor job it is 1, here it is shared over 4 processors (Appears from 'shared processor' in GUI)
%mem=4GB: Indicates the amount of dynamic memory allocated (Appears from 'memory Limit' in GUI)
%chk=file.chk: Name of checkpoint file (Appears from 'Checkpoint File' in GUI)
 # p opt freq hf/6-31+g(d,p) geom=connectivity pop=Full: Here keywords to be used. This part specifies which type of calculation user wants to do. This part will be discussed in the next section in detail

H2O→ Title Section
0 1 (1st indicates charge (0) and second indicates multiplicity (1))
O 0.27861445 0.40662650 0.00000000 (Cartesian coordinate of Oxygen)
H 1.23861445 0.40662650 0.00000000 (Cartesian coordinate of Oxygen)
H -0.04184014 1.31156233 0.00000000 (Cartesian coordinate of Oxygen)

1 2 1.0 3 1.0 (This section indicates connectivity of atoms between other atoms and type of bond. 1st atom O is connected with 2nd and 3rd by single bond, thus 1 2 1.0 2 3 1.0)
2
3

Fig. (10.8). Gaussian input file for water molecule.

II. TO FIND OUT ENERGY OF ETHANE AT VARIOUS CONFORMATIONS

Theory

In an ethane molecule, two carbon atoms are bonded by a single bond; hence, they can rotate freely in any direction. As a result, a huge number of conformations of ethane are anticipated. Interestingly the energies of the different conformation are not the same. Those differences in energies are arising due to the interaction between the hydrogen atoms attached with C_1 and C_2.

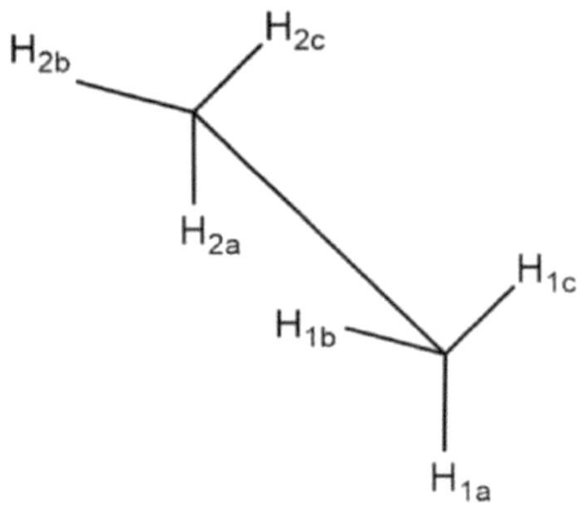

Fig. (10.9). Eclipse conformation (Sawhorse projection) of ethane.

Experimentally, the calculation of this difference in energy for various conformations is impossible. However, using the quantum chemical method we can easily calculate the energy of various conformations of ethane. Herein, we have adopted the basic quantum chemical method, *i.e.,* the Hartree-Fock method. The Hartree-Fock method is briefly outlined below:

Short Description of the Hartree-Fock (HF) Method

In the Hartree-Fock method, the many-electron system is approximated to be many one-electron systems. The Hamiltonian consists of two parts, *viz*; kinetic energy part or one electron term, which takes care of the motion of the electron as well as the electron-nucleus interaction, and another part is potential energy term, which is further bifurcated into two terms, Coulomb interaction potential and exchange potential. The coulombic interaction arises due to the interaction (or repulsion) between two electrons. In practice, the coulombic form is considered as interaction between electron one and rest (n-1) electrons in a spherically average manner while the exchange term has no classical mechanical analogue. This arises due to the exchange of electrons one with other (n-1) electrons. In the Hartree-Fock method exchange interaction is considered for an electron with parallel spin. The wavefunction of "n" electrons generally represent using Slater's determinant as given in Equation (10.16).

$$\Psi\ (X_1X_2X_3 \ldots . X_n) = \begin{vmatrix} X_1(x_1)X_1(x_2)X_1(x_3) \ldots \ldots \ldots \ldots . X_1(x_n) \\ X_2(x_1)X_2(x_2)X_2(x_3) \ldots \ldots \ldots \ldots . X_2(x_n) \\ X_3(x_1)X_3(x_2)X_3(x_3) \ldots \ldots \ldots \ldots . X_3(x_n) \\ \ldots \ldots \ldots \ldots \ldots \ldots \ldots \ldots \ldots \ldots \ldots \ldots \ldots \ldots \ldots \ldots \ldots \\ \ldots \ldots \ldots \ldots \ldots \ldots \ldots \ldots \ldots \ldots \ldots \ldots \ldots \ldots \ldots \ldots \ldots . \\ X_n(x_1)X_n(x_2)X_n(x_3) \ldots \ldots \ldots \ldots . X_n(x_n) \end{vmatrix} \quad \textbf{(10.16)}$$

X_i = spin orbital x_i = coordinates of orbital

In short hand notation, Equation (10.16) can represent as

$$\Psi\ (x_1, x_2, x_3 \ldots x_n) = \lceil x_1x_2x_3 \ldots \ldots x_n \rangle \quad \textbf{(10.17)}$$

The Slater determinant is solved by using the variational method. In the variational method the form of the Hartree-Fock method is represented in Equation (10.18):

$$\text{H}\Psi = \text{E}\Psi \quad \textbf{(10.18)}$$

$$\textstyle\sum_{i=1}^{N}\langle X_i|n|X_i\rangle + \sum_{i>j}\left[\langle X_iX_j|J|X_iX_j\rangle - \langle X_iX_j|K|X_jX_i\rangle\right] = \text{E}\Psi \quad \textbf{(10.19)}$$

However, this equation cannot be solved easily, in order to solve this equation Roothan developed a unique method, which is popularly known as the Roothan-Hartree-Fock method. Expressed as:

$$\mathbf{FC{=}EC\varepsilon} \quad \textbf{(10.20)}$$

Here, **F**, **C**, **S**, and ε denote the Fock matrix, co-efficient matrix, overlap, and the orbital energies respectively, which appear in the diagonal term. Notably, the Fock matrix is of order i**X**j, As the **C** matrix is present on both sides of the equation, we can't solve this equation directly. Instead of this, we have to use the iterative method. Hence, this method is known as Self-Consistence Field (SCF) method. Further to solve the equation correctly, we need to guess the wavefunction by introducing basic sets.

Unlike hydrogen atoms, we cannot solve for energy in many-electron systems. Instead, we need to adopt a basis set, which is based on the postulate that a particular wave function can be represented as a linear combination of other wave functions, *i.e.*,

$$\boldsymbol{\Psi}_i = \textstyle\sum_j \mathbf{c}_i\, \boldsymbol{\varphi}_j \quad \textbf{(10.21)}$$

Here, $\mathbf{c_i}$ and $\boldsymbol{\varphi_j}$ are known as coefficient and atomic orbitals.

In order to get exact energy, theoretically, 'i' should be up to infinity, but in practice that is impossible. Instead, we need to take a large basis set.

In this context presenting modified Gaussian type orbital showed computational efficiency. Moreover, these are as good as Slaters-type orbitals. Hence, for this purpose, we will use 6-31G basis sets. Here 6- Gaussian type orbitals of individual atoms, whereas 3 and 1 Gaussian type orbitals are used to represent the valence electronic orbitals.

For example, for carbon atom 6 Gaussian type orbitals are used to represent the 1s orbital and 3 Gaussian type orbitals are used to represent 2s orbital and 1 Gaussian type orbitals are used to represent 3p orbital.

Procedure

1) The structure of the ethane molecule was drawn using Gauss View software.

2) To calculate the energy of various conformations of ethane, we have chosen the HF method and 6-31 G basis set.

3) Single point energy option has opted in the ground state. For that just choose energy.

4) Memory limit is chosen to be 1GB as we are using a single processor. If the shared processor is increased memory limit should also be increased accordingly. The calculations are performed in the gas phase and the rest of the part is taken as default as provided by the software when the resulting file is saved.

5) Similarly, various other conformations are built up with the difference in dihedral angle with ±15° are constructed, and step 2-5 are repeated for all the structures.

6) Single point energies of all these structures are noted from the output file (*filename*.log file) by using the keyword HF.

7) A plot is done using MS Excel where energy and dihedral angle along the Y and X-axis are considered.

8) Several minima are obtained in the plot. Among them, the minima having the largest absolute value of energy will correspond to the global minima (optimized dihedral angle).

Result

Plot dihedral angle *vs*. Energy. There will be more than one minimum point in the plot. The minima corresponding to the highest modulus of energy value (most negative) will correspond to the globally minimum.

Dihedral angle in most stable conformation is = ...°.

The energy corresponding to .... Dihedral is ...... a.u.

The plot of the variation of energy with respect to dihedral angle change is shown in Fig. (**10.10**).

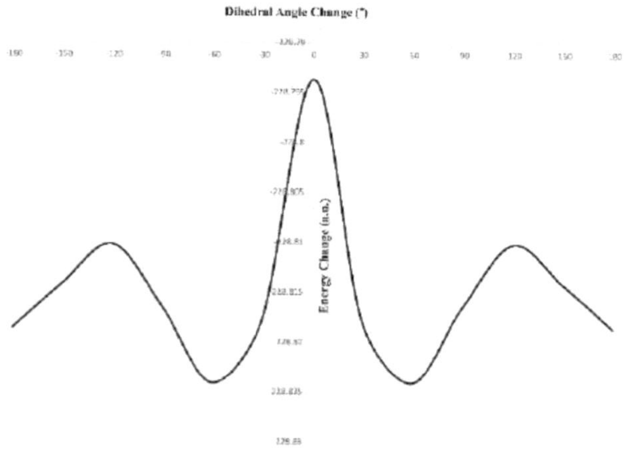

Fig. (10.10). The plot of the variation of energy with respect to dihedral angle change in ethane.

III. TO FIND OUT THE ENERGY OF 1, 2-DIHYDROXY ETHANE AT VARIOUS CONFORMATIONS

Theory

In the 1, 2-dihydroxy ethanol molecule, two carbon atoms are bonded by a single bond resulting they can rotate along with the bond freely. Thus, we can have a large number of configurations. Interestingly, the energy of different conformations is not the same. The difference in the energies arises due to the interaction between the various groups attached with two different C–atoms. Since in this molecule two hydroxy groups are present, we can expect intramolecular hydrogen bonding in this system.

Method

HF method is used. Details of the HF method and basis set are given in experiment II.

Software Used

Gaussian09 Revision D 01 and Gauss View 5.0.9.

Procedure

1) The structure of the 1, 2-dihydroxy ethane molecule was drawn using Gauss view 5.0.9 software.

2) To calculate the energy at various conformations, we have preferred the HF method and 6-31G basis set.

3) The memory limit is chosen to be1GB since we are using a single processor.

4) All the calculations are performed in the gas phase.

5) Rest of the things are taken as default as provided by software and the resulting file is saved.

6) The output files are further opened for calculation of energy using Gauss View 5.0.9 software.

7) Various other conformations with different dihedral angles HO-C1-C2-OH by the dihedral angle of (±) 15° are constructed and step 2) to step 6) are repeated for all the conformations.

8) The energies of individual structures are noted from the output files by using the keyword RHF.

9) A plot is drawn using MS-Excel where energy is in the Y-axis and dihedral angle in the X-axis. The minimum represents the global minima of all the conformations and this particular conformation is the most stable conformation of 1,2-dihydroxy ethane.

Result

Plot dihedral angle *Vs*. energy. There will be more than one minimum point in the plot. The minima corresponding to the highest modulus of energy value (most negative) will correspond to the global minimum.

Dihedral angle in most stable conformation is = ...°.

The energy corresponding to .... dihedral angle is ...... a.u.

Compare the dihedral angles corresponding to minima for ethane and 1, 2-dihydroxy ethane.

The plot of the variation of energy with respect to dihedral angle change is shown in Fig. (**10.11**).

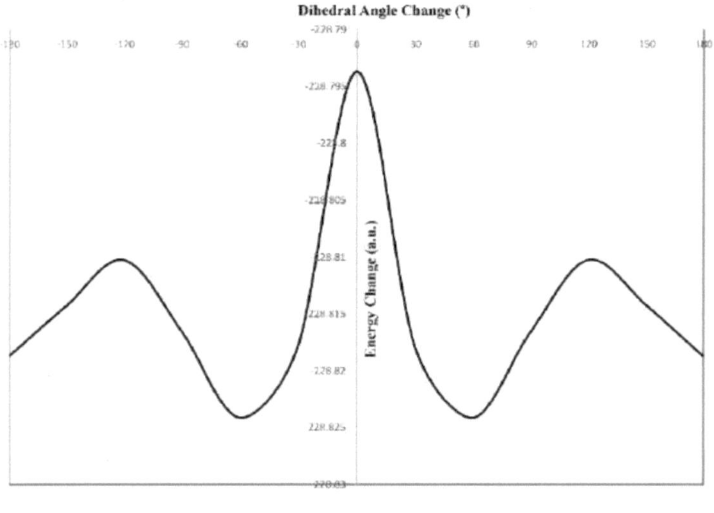

Fig. (10.11). The plot of the variation of energy with respect to dihedral angle change in 1,2-dihydroxy ethane.

IV. SCAN AROUND A SINGLE BOND AND FIND OUT THE MINIMUM ENERGY OF A DIATOMIC MOLECULE

Theory

For a diatomic molecule, the only way to find out the minimum energy of a system is to find out the change in the bond length.

In a diatomic molecule, it is easiest to do by using an automated method. In this automated method, a scan around a particular bond, bond angle, or dihedral angle to find local minimum energy by using single-point energy, is carried out. In the scan method, a rigid potential energy surface (PES) scan is performed through the calculation of single-point energy evaluations over a rectangular grid relating to selected internal coordinates. Herein, the number of steps and step size for each variable is specified on the variable definition lines, following the variable's initial value.

In this work, we have tossed N_2 molecule to find out the energy at different bond lengths which vary from 1Å - 4 Å with an interval of 0.1 Å, and find out how the bond energy changes with change in bond distance. A plot of bond distance *Vs.* energy is shown below. The minimum energy and corresponding bond length correspond to optimized energy and optimized bond length.

Method

HF method is used. Details of the HF method and basis set are written in experiment II.

Software Used

Gaussian09 Revision D 01 and Gauss View 5.0.9

Procedure

1) The N_2 molecule is built up first. To incorporate the scan keyword, we have done the following: from the edit section, the redundant coordinates option was chosen and from there, a bond between 1 and 2 had been selected.

2) The coordinates and the scan coordinates were marked as follows:

Coordinates	**1**	**2**
Scan Coordinate	**Take 30 steps of size 0.1A°**	

3) In calculate section, the redundant coordinate scan was chosen. Like the previous system, the Hartree Fock method along with the 6-31G basis set has opted.

4) The output file was opened in Gauss View and the scan option was chosen from the result section.

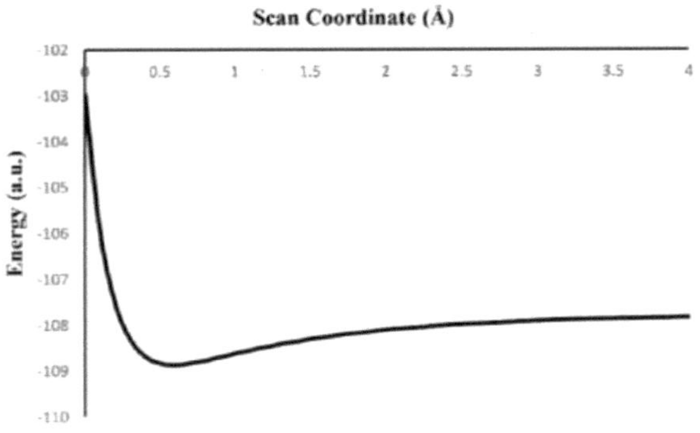

Fig. (10.12). Plot of Scan Coordinate *vs.* Total Energy (a.u.) for nitrogen molecule.

Result

The local minimum energy of N_2 molecule is =...... Hartree (a.u.).

The bond length at minimum energy or equilibrium bond length is = ... Å.

Bond dissociation energy = ... a.u.

The plot of the Scan coordinate *vs*. the total energy is similar to Fig. (**10.12**).

V. FIND OUT OPTIMIZED GEOMETRY OF WATER MOLECULE

Geometry Optimization

Optimization of molecular geometry is a very important task in quantum chemistry. In the Gaussian 09 and other standard software, the optimized geometry of a molecule can be calculated efficiently. Atomic arrangements within the molecule will be adjusted by geometric optimization until the minimum energy is achieved. In other words, the point is known as the minima on the potential energy surface. This is the lowest position on the potential energy surface. (Fig. **10.13**) shows a typical flow chart explaining the geometry optimization process. The optimized geometry of the water molecule is shown in Fig. (**10.14**).

Frequency Calculations

Vibrational spectra of molecules in their ground and excited states can be computed by employing the Gaussian program. Frequency calculations are valid only at the stationary points on the potential energy surface, for this reason, it is necessary to run a geometry optimization before executing a frequency calculation. In addition, to predict the frequencies and intensities of infrared spectra, the program can also describe the displacements of the molecule as it undergoes normal modes of vibrations. Molecular frequencies depend on the second derivative of energy with respect to the position of nuclei. The code for accurate analytic second derivatives of energy is available in the Hartree-Fock and various post-Hartree-Fock methods. Harmonic force fields derived from quantum mechanics are widely used at present for the calculation of frequencies and the normal modes of vibration. This opened the way to calculate the frequencies and intensities of spectral bands with a minimum degree of arbitrariness and finding the rational explanations for a number of chemical and physical properties of substances.

The geometry optimization process includes two different optimisations. 1) Optimize the electronic position and 2) Optimize the nuclear coordinates. Both the process involves iterative techniques. The conditions for the nuclear iteration process include the convergence of four criteria. These are maximum force, Root Mean Square (RMS) force, maximum displacement, and RMS displacement. For individual geometry, the output file is printed. When the convergence criteria are fulfilled, gaussian calculates the frequencies for each normal mode of vibration. The absence of non-negative frequency is indicative of the achievement of true minima, but the presence of one negative frequency indicates a transition state is attained. Thus, completion of the iterative process at a particular geometry should meet their convergence criteria and the energy of the system is printed.

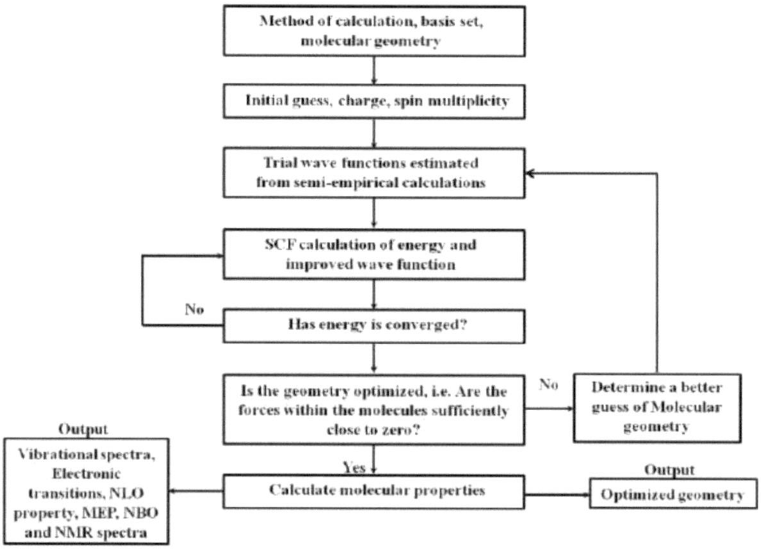

Fig. (10.13). Flowchart illustrating the steps involved in quantum chemical calculations of molecular structures and properties.

Method

HF method is used. Details of the HF method and basis set are written in experiment II.

Software Used

Gaussian09 Revision D 01 and Gauss View 5.0.9.

Procedure

1) Build the structure of water molecules using Gauss View 5.0.9 software package.

2) Now go to the calculation scheme and opt for the optimization and frequency.

3) The HF method along with the 6-31G basis set have opted.

4) Save the structure and run for optimization.

Result

Individual Optimum Bond Lengths = …… Å.

Optimum Bond Angle =…..°.

Optimum Dihedral Angle (Not applicable for water molecule).

Frequencies = ….

Energy = …. a.u.

Represent the output for this particular molecule as follows:

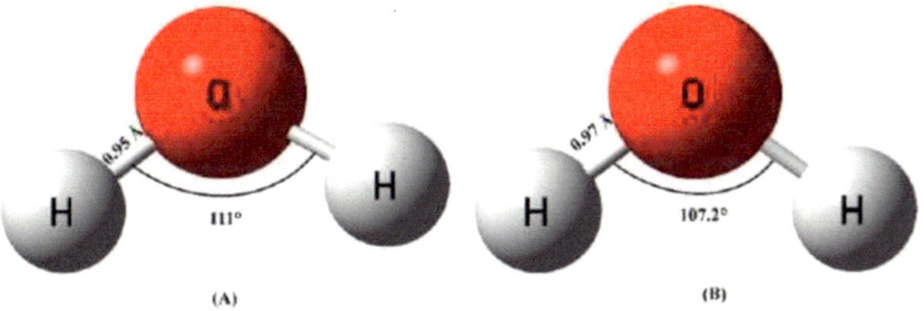

Fig. (10.14). Optimised geometry of water molecule. **(A)** Before optimisation, **(B)** After optimisation.

VI. EFFECT OF BASIS SET ON THE OPTIMISATION

Theory

The description of the 6-31 G basis set is already described in Experiment - 1, but the rest of the basis sets will be described a little bit in this experiment.

STO-3G Basis Set

Developed by John F. Pople. He used the "SP" shell which represents the s and p function. AO (Atomic orbital) is represented by 3 Gaussians chosen to mimic the behaviour of an STO (Slater Types Orbital).

6-311G Basis Set

It is a split value triple–zeta basis. It adds one G to (Gaussian Type orbital) to 6-31G.

3-21G Basis Set

Core orbital is a sum of 3 Gaussians. 2s and 2p orbitals are split into two parts: the inner part is a sum of two Gaussians outer part is one Gaussian. When an atom is in an anion or an excited state, the loosely bonded electrons, which are responsible for the energy in the tail of the wave function became much more important. This is where computational scientists use diffuse function. Diffuse basis sets are represented by the '+' signs. One '+' means we are accounting 'p' orbitals which '++' focuses both on p and d orbitals.

Method

HF method is used. Details of the HF method and basis set are written in experiment II.

Software Used

Gaussian09 Revision D 01 and Gauss View 5.0.9.

Procedure

1) The molecular structure of nitrogen molecule is built using Gauss view 5.0.9 software.

2) The structure was optimized using a 6-31G basis set and the HF method is used.

3) Step 1 and Step 2 are repeated for various basis sets such as 3-21G, 6-311G, 6-311G+(d, p), 6-311G+(d) keeping the quantum chemical method unchanged.

4) The optimized bond length and minimum energy for different basis set is noted down.

Result

All the energies and bond lengths with respect to the basis set are noted in Table **10.1**.

Table. 10.1. Variation of bond length and minimum energy of N_2 molecule using different basis sets.

Basis Set	Optimum Bond length (Å)	Minimum Energy (a. u.)
6-31 G		
3.21G		
6-311G		
6-311G + (d)		
6-311G+ (d, p)		

VII. N_2 AND CO BOTH ARE ISOELECTRONIC BUT N_2 IS INERT AND CO IS REACTIVE: A COMPUTATIONAL INSIGHT

Theory

It is a very common undergraduate question that although nitrogen molecule and CO molecule both contain the same number of electrons (isoelectronic), we observe nitrogen molecule is inert, but CO molecule is highly reactive. The answer can be addressed by using their molecular orbital diagram, which says that the electrons pairs of the Highest Occupied Molecular Orbital (HOMO) of N_2 molecule reside in bonding orbital (σ_g ($2p_z$)), whereas the same for CO belongs to non-bonding σ (2pz)) orbital. As a result, CO can show Lewis base behaviour by donating the lone pair of electrons and it is highly reactive. In this study using the quantum chemical approach, we shall explain the reason. When we optimize the geometries of both the molecules and analyse their HOMO, then we shall see that the electrons are uniformly distributed over two nitrogen atoms in the N_2 molecule and this orbital belongs to gerade symmetry, but the HOMO electrons density is mainly localized on the carbon centre. Hence, the HOMO of CO is assigned as non-bonding orbital, which is mainly contributed from the $2p_z$ atomic orbital of CO (Assuming z-axis as the bonding axis). Thus, despite both N_2 and CO has 14 electrons, but CO shows reactivity, but another is chemically inert. The MO electron density plot for both N_2 and are shown in Fig. (**10.15**).

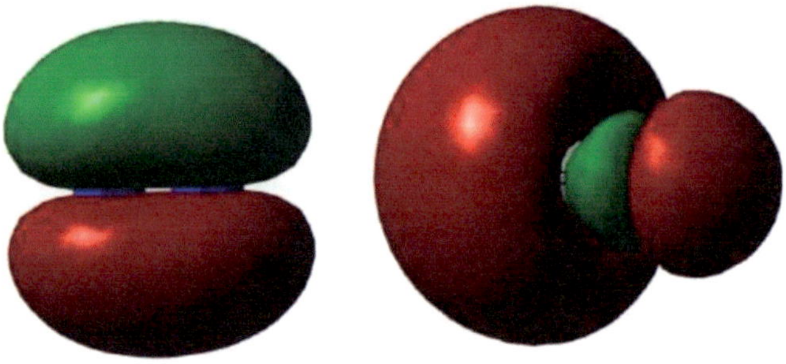

(A) (B)

Fig. (10.15). Electron density plot for **(A)** N_2 and **(B)** CO molecule with isosurface value of 0.02 and the green magenta colour represent alpha and beta electrons respectively.

Method

HF method is used. Details of the HF method and basis set are written in experiment II.

Software Used

Gaussian09 Revision D 01 and Gauss View 5.0.9.

Procedure

1) The structure of both N_2 is built using the Gauss view 5.0.9 software package.

2) Geometry of N_2 is optimized in the gas phase using HF level of theory and a 6-31 G basis set was employed to describe the AOs of the atoms. Also, ensure true minima have been achieved by the absence of non-negative frequency.

3) The checkpoint file (filename.chk) of the optimized structure is opened in the Gauss View software package.

4) Take the electron density plot for the molecule. The electron density can be obtained as follows:

(A) Go to the Results section in the Gauss View window.

(B) Select Surface/Contours.

(C) Select New Cube from Cube Actions and opt for the HOMO orbital.

(D) Finally select New Surface from Surface actions. The MO density plot will be opened with a default isosurface value of 0.02.

5) Repeat Step 1 to Step 4 for the CO molecule.

6) Compare the electron density plots for both nitrogen and carbon monoxide.

Result

The electron density plots indicate the HOMO of N_2 belongs to the sigma bonding (σ_g) orbital and the same for CO belongs to the non-bonding (σ_{2p_x}) orbital.

Comment

Repeat the calculations with correlated methods, like, MP2, CCSD, DFT, *etc.*, and check how the orbital energies are changed.

VIII. TO CALCULATE HYDROGEN BONDING ENERGY FOR WATER MOLECULE

Theory

Hydrogen bonding (H-bonding) plays a vital role in the formation of living beings. However, H-bonding belongs to neither a co-valent nor an ionic-bonding, rather it is a kind of electrostatic interaction that takes place between covalently bonded H and high electronegative atoms like N, O, F.

There are several direct pieces of shreds of evidence towards the formation of H-bonding in various systems. *e.g.,* H_2O is a liquid but H_2S is gas, only because of H-bonding in water. Using IR spectroscopy, we can detect the formation of H-bonding by lowering the frequency of that particular bond. In the quantum chemical calculations, we will calculate the H-bond energy by the direct subtraction method. Here, we have coined H_2O dimer to study the H-bonding. The H-bond energy can be expressed as:

$$E_{H-bond} = E_{(water-dimer)} \times 2E_{(water)} \quad \textbf{(10.21)}$$

Here, $E_{H\text{-}bond}$ indicates H-bond formation energy. A negative value of E is indicative of the favourable formation of the hydrogen bond. Whereas, the opposite symbol indicates not a formation of H-bond. The first term on the right-hand side of Equation (10.21) indicates the energy of hydrogen-bonded water dimer and $E_{(water)}$ represents the energy of a single water molecule, respectively.

To get H-bonding energy very accurately, one needs to compute the Basis set Super Position Error (BSSE) by using the counterpoise correction of Boys and Bernardi [15]. It is quite convincing to calculate the interaction energy between the two monomers of water as a difference of energies of water dimer and sum of two water molecules. To calculate BSSE, we use the keyword "counterpoise=2".

Method

HF method is used. A detail of the HF method and basis set is written in experiment II.

Software Used

Gaussian09 Revision D 01 and Gauss View 5.0.9.

Procedure

1) Structures of the water molecule and water-dimer were built in the Gauss view 5.09 software package. Since the bond distance for H-bonding is 2Å. Hence, we should keep the same in mind during the construction of the H_2O dimer.

2) Structures of H_2O molecule and H_2O dimer are optimized at HF/6-31G level of theory. In this context, while optimizing the water dimer include the BSSE factor by introducing "counterpoise=2". The necessary keywords in the input are marked by bold text:

opt freq hf/6-31g geom=connectivity counterpoise=2

BSSE Corrected Input for Water Dimer

```
0 1
O    -3.26694001  -2.13926196   0.05565222  1
H    -2.30694001  -2.13926196   0.05565222  1
H    -3.58739460  -1.23432613   0.05565222  1
O    -0.35642969  -1.93006383  -0.33389320  2
H     0.60357031  -1.93006383  -0.33389320  2
H    -0.67688428  -1.02512800  -0.33389320  2
```

3) The energy of the 2 entities is noted down. Notably, Take the BSSE corrected energy for the dimer.

4) The interaction energy due to the formation of the H-bond is calculated using Equation (10.21).

5) Any drop in OH-bond frequency indicates that the particular H atom is involved in H-bond formation.

Results

Energy of water monomer = … a.u.

Energy of water dimer (BSSE corrected) = … a.u.

BSSE = …. a.u.

Complexation Energy = … kJ/mol

H-bond energy in water = (***) a. u. = (***) ×627.50 kcal/mol = (***)×2625.5 kJ/mol.

Comment

Using the Non-Covalent Interaction (NCI) plot the hydrogen bonding can be observed and also visualize the hydrogen bond formation using the VMD software package with the HBONDS plugin.

IX. TO FIND (A) OPTIMIZED GEOMETRY, (B) THE EFFECT OF HYDROGEN BONDING AND (C) COMPUTATION OF IR-FREQUENCIES

Theory

Hydrogen bonding (H-bonding) plays a vital role in the formation of the living being. However, H-bonding belongs to neither a co-valent nor an ionic-bonding, rather it is a kind of electrostatic interaction that takes place between covalently bonded H and high electronegative atoms like N, O, F.

There are several direct pieces of shreds of evidence towards the formation of H-bonding in various systems. *e.g.,* water is a liquid but H_2S is gas, only because of H-bonding in water. Using IR spectroscopy, we can detect the formation of H-bonding by lowering the frequency of the particular bonds. Based on the formation of H-bonding, we can classify the H-bonding as intermolecular H-bonding and intramolecular H-bonding. In simple quantum chemistry calculation, we can calculate the intermolecular H-bond energy by the direct subtraction method. Let A and B are two molecules that form intermolecular H-bonding. The H-bond energy can be expressed as:

$$E_{H\text{-bond}} = E_{AB} - (E_A + E_B) \quad \textbf{(10.22)}$$

Here, $E_{H\text{-bond}}$ indicates H-bond formation energy. A negative value of E is indicative of the favourable formation of the hydrogen bond. Whereas, the opposite symbol indicates the non-formation of the H-bond. The first term on the right-hand side of Equation (10.22) indicates the energy of hydrogen-bonded AB and E_A and E_B represent the energy of A and B molecule, respectively.

In order to get H-bonding energy very accurately, one needs to compute the Basis set Super Position Error (BSSE) by using the counterpoise correction of Boys and Bernardi [15]. It is quite convincing to calculate the interaction energy between the two monomers of water as a difference of energies of water dimer and sum of two water molecules. To calculate BSSE, we use the keyword "counterpoise=2".

However, the computation of intramolecular H-bonding is not so straightforward. A good approach to realizing H-bond formation is by comparing the IR frequencies. To do so, we shall optimize the structures twice. Once arrange the molecules in such a way, that intramolecular H-bond cannot be formed optimize the structures and optimize the molecules in such orientation, such that intramolecular H-bond can be formed. Study the computed IR frequency data for both the optimized structures from the results → vibrations → IR frequency, and note the change in IR frequency of that particular normal mode of vibration.

To perform this, we have chosen (A) *p*-hydroxy benzaldehyde; (B) *o*-hydroxy benzaldehyde for intermolecular H-bonding and intramolecular H-bonding.

Method

HF method is used. Details of the HF method and the basis set are written in experiment II.

Software Used

Gaussian09 Revision D 01 and Gauss View 5.0.9.

Procedure

1) All three structures under consideration are built-in gauss view 5.0.9 software packages and the structures are shown in Fig. (**10.16**). The monomers are optimized at the HF/6-31 (G) level of theory.

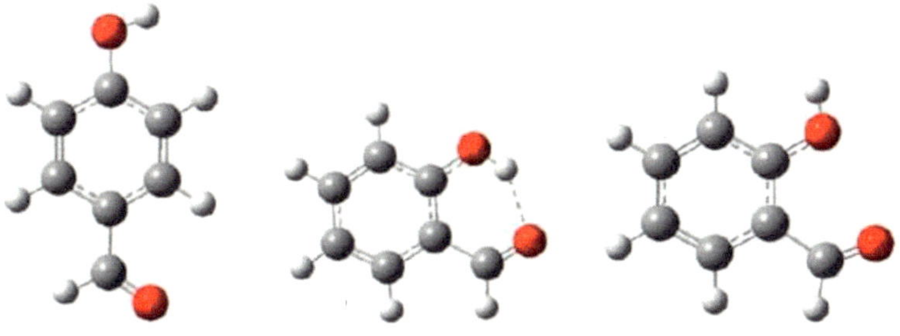

(A) (B) (C)

Fig. (10.16). Optimized structures of **(A)** *p*-hydroxy benzaldehyde; **(B)** *o*-hydroxy benzaldehyde (phenolic H and carbonyl O are in proximity); **(C)** *o*-hydroxy benzaldehyde (phenolic H and carbonyl O are in the far distance).

2) For *p*-hydroxy benzaldehyde dimer we optimize the structure. BSSE correction was incorporated as mentioned in the previous experiment and the same level of theory is applied.

3) The *o*-hydroxy benzaldehyde can have two different orientations as shown in (B) and (C) in Fig. (**10.16**). Both the structures are optimized using the same level of theory, but BSSE correction is not necessary.

4) The interaction energy due to the formation of intermolecular H-bond is calculated using Equation (10.22).

5) Note the frequencies and IR spectra for all the molecules in monomer, dimer, and intramolecular H-bonded situations. Also, compare the loss or gain in frequency for *o*-hydroxy benzaldehyde.

Results

Report the energies of all three isomers in Table **10.2**. Take relative energies by taking the difference between a particular molecule and the most stable energy. Note down the normal modes of vibrations in *o*-hydroxy benzaldehyde and report (Table **10.3**).

Table 10.2. Energies of monomers all three isomers.

Molecule	Energy (a.u)
p-hydroxy benzaldehyde	
o-hydroxy benzaldehyde (H and O are proximal)	
o-hydroxy benzaldehyde (H and O are non-proximal)	

BSSE corrected energy of *p*-hydroxy benzaldehyde = … a.u.

BSSE = …. a.u.

Complexation Energy = … kJ/mol

Hydrogen bond energy in *p*-hydroxy benzaldehyde = (***) a. u. = (***) x 627.50 kcal/mol = (***) x 2625.5 kJ/mol

Table 10.3. Computed Normal Modes of vibrations in *o*-hydroxy benzaldehyde.

Frequency	Atoms involved in (B)	Frequency	Atoms Involved in (C)	Difference in Frequency
ν_1				
ν_1				
ν_1				

X. HYDROGEN BOND ENERGY IN DNA BASE PAIRS

Theory

The H-bonding belongs to neither a co-valent nor an ionic-bonding, rather it is a kind of electrostatic interaction that takes place between covalently bonded H and high electronegative atoms like N, O, F and hydrogen bonding (H-bonding) play a vital role towards the formation of the living being, *e.g.*, DNA consists of four base pairs, *viz.* Adenine, Guanine, Cytosine, and thymine. In the DNA they cannot form base pairs randomly. Instead, they form in a particular fashion. Like, Adenine (A) forms pair with Thymine (T) through a couple of hydrogen bonding and Cytosine (C) pairs up with Guanine (G) *via* three hydrogen bonding. These bonds are shown in Fig. **(10.17)**. Using quantum chemical calculation, we can calculate the intermolecular H-bond energy of these base pairs by the direct subtraction method. Let A and B are two molecules that form intermolecular H-bonding. The H-bond energy can be expressed as:

$$E_{H\text{-bond}} = E_{AB} - (E_A + E_B) \qquad \textbf{(10.22)}$$

Here, $E_{H\text{-bond}}$ indicates H-bond formation energy. A negative value of E is indicative of the favourable formation of the hydrogen bond. Whereas, the opposite symbol indicates not the formation of the H-bond. The first term on the right-hand side of Equation (10.22) indicates the energy of hydrogen-bonded AB and E_A and E_B represent the energy of A and B molecule, respectively.

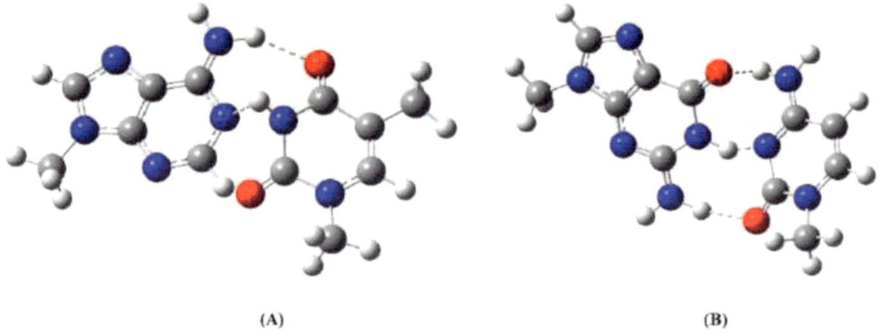

Fig. (10.17). Hydrogen bonded base pairs **(A)** Adenine **(A)**-Thymine (T); **(B)** Guanine (G)-Cytosine **(C)**.

In order to get H-bonding energy very accurately, one needs to compute the Basis set Super Position Error (BSSE) by using the counterpoise correction of Boys and Bernardi [15]. It is quite convincing to calculate the interaction energy between the two monomers of water as a difference of energies of water dimer and sum of two water molecules. To calculate BSSE, we use the keyword "counterpoise = 2".

Method

HF method is used. Details of the HF method and basis set are written in experiment II.

Software Used

Gaussian09 Revision D 01 and Gauss View 5.0.9.

Procedure

1) All the four bases (A, T, G, and C) under consideration are built-in gauss view 5.0.9 software packages, and the structures are shown in Fig. (**10.18**). The monomers are optimized at the HF/6-31 (G) level of theory.

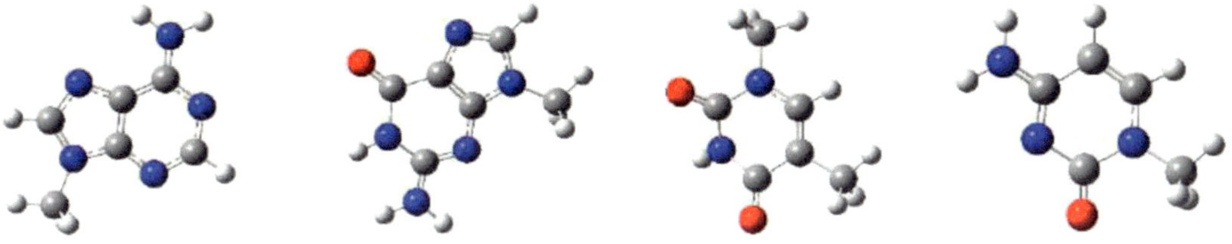

(A) (B) (C) (D)

Fig. (10.18). Optimized structures of **(A)** Adenine; **(B)** Guanine; (**C**)Thymine; **(D)** Cytosine.

2) A-T and G-C pairs are optimized by introducing hydrogen bonds as shown in Fig. **(10.17)** with the BSSE correction.

3) The interaction energy due to the formation of intermolecular H-bond is calculated using Equation (10.22).

Results

Report the energies of the individual bases (Table **10.4**) as well as the base pairs (A-T and G-C). Take the relative energies by taking the difference between a particular molecule and the most stable energy.

Table 10.4. Energies of monomers all the bases.

Molecule	Energy (a.u)
Adenine	
Guanine	
Thymine	
Cytosine	

BSSE corrected energy of A-T base pair = … a.u.

BSSE = …. a.u.

Complexation Energy = … kJ/mol.

Hydrogen bond energy in A-T base pair = (***) a.u. = .

(***) x 627.50 kcal/mol = (***) x 2625.5 kJ/mol.

BSSE corrected energy of G-C base pair = ... a.u.

BSSE = ... a.u.

Complexation Energy = ... kJ/mol.

Hydrogen bond energy in G-C base pair = ... a.u.

(***) x 627.50 kcal/mol = (***) x 2625.5 kJ/mol.

XI. GEOMETRY OPTIMIZATION OF $[Co(H_2O)_6]^{2+}$ COMPLEX USING RESTRICTED OPEN HARTREE-FOCK (ROHF) AND UNRESTRICTED HARTREE-FOCK (UHF) METHOD

Theory

Geometry optimization of non-transition elements is easier by using a Gaussian basis set. However, the same for compounds having transition metal or heavier elements is not that easy as it consists of many electrons on an atom. In order to reduce the computational cost, we have to apply effective core potential (ECP) for these elements and hence, we need to apply different basis set on these elements. For heavier elements we shall use numerical basis sets with ECP and the Pople's basis set can be used for lighter elements. In the ECP method, we use approximate potential energy from all core-potential rather than all electrons which gives results from the computational bottleneck without much loss of accuracy in the calculation.

In this practical, we shall optimise the structure of $[Co(H_2O)_6]^{2+}$. Here the H_2O molecules are occupying the octahedral site of the Co(II) ion. We shall apply a 6-31G basis set on H and O. Los Alamos National Laboratory 2 Double-Zeta (LanL2DZ) basis set will be used for Co (II) ion.

Further, since the complex contains unpaired electrons with minimum spin state doublet. Hence, we cannot apply the Restricted Hartree-Fock (RHF) method, which is applicable for closed-shell (When all electrons are paired up) molecules. Instead, we have to use Restricted Open Hartree-Fock (ROHF) method or the Unrestricted Hartree-Fock (UHF) method. In the ROHF method, the single electron is considered separately, but the paired electrons are treated as RHF method, but in

UHF different molecular orbitals are considered for the electrons having a different spin. Since the HF method considers the correlation between parallel electrons, hence it is observed that the degeneracy of spin orbitals having different spin changes. The change will be dictated by the unpaired electron, *e.g.,* if the unpaired electron has α spin, the energy of other α spin electrons will be lesser than the β spin electron (having more negative value).

Method

HF method is used. A detail of the HF method and basis set is written in experiment II.

Software Used

Gaussian09 Revision D 01 and Gauss View 5.0.9.

Procedure

1) The structure of the complex ion was built up and the file was saved.

2) File was then reopened and go to the "Result" section and select "View file".

3) Since the molecule contains unpaired electrons with a minimum spin state is a doublet. Hence, we cannot apply the restricted Hartree-Fock method. In this case, we can apply the Restricted open Hartree-Fock (ROHF) method or unrestricted Hartree-Fock (UHF) method.

4) The file was edited and saved again as follows control key:

" # opt freq UHF/genecp"

Go to the bottom of the file and write the following keywords:

Leave single line spacing

H,O, 0

6-31+g

****

Leave single line spacing

Co, 0

LanL2DZ

****

Press "enter"

Co, O

LanL2DZ

Leave single line spacing

5) Optimized the structure using the UHF method. After optimization check the frequency of the output file to ensure true minima has been achieved by the absence of negative frequency.

6) Repeat the geometry optimization of the same complex by replacing UHF with ROHF. After optimization check the frequency of the output file to ensure true minima has been achieved by the absence of negative frequency.

7) After optimization compare the energies and the orbital energies of both structures.

Result

Report the results (i) Lowest three frequencies of the complex optimized in different methods (ii) Optimized Energy of the complex optimized in different methods and (iii) Orbital Energies (Highest five occupied orbitals) of the complex optimized in different methods in Table **10.5**, **10.6** and **10.7** respectively.

Table 10.5. Lowest three frequencies of the complex optimized in different methods.

Frequency Value	UHF	ROHF
ν_1		
ν_2		
ν_3		

Table 10.6. Optimized Energy of the complex optimized in different methods.

UHF Energy (a.u.)	ROHF Energy (a.u.)	Energy Difference (a.u.) (UHF Energy - ROHF Energy)	Energy Difference (Kcal/mol)

Table 10.7. Orbital Energies (Highest five occupied orbitals) of the complex optimized in different methods.

Orbital	Unrestricted (UHF)		Restricted open (ROHF)	
	α-electron	β- electron	α- electron	β- electron
HOMO				
HOMO- 1				
HOMO- 2				
HOMO- 3				
HOMO-4				

XII. EFFECT OF CORRELATION IN QUANTUM CHEMISTRY

Theory

Although the Hartree-Fock method is 95%-98% accurate for the ground state calculations, chemists are interested in energy difference. While calculating the energy difference, this small percentage of error leads to a large error value. Other than this, the HF method has some severe flaws. These are:

1) HF method cannot account for the instantaneous electron repulsion.

2) HF method nicely explains the ground state energy and properties of molecules, but it cannot explain the excited state and quasi degenerate states of molecules.

3) In the HF method correlation between electrons of parallel spin is considered, but not the opposite spin.

HF method fails to explain several important phenomena in chemistry, *e.g.*, the dissociation of molecules in open-shell fragments, which arises due to neglecting the correlation between electrons of anti-parallel spin, ionization potential of a nitrogen molecule. Thus, these errors arise due to correlation. It is defined as the difference between exact energy and Hartree-Fock energy and expressed as:

$$E_{corr} = E_{exact} - \varepsilon \qquad \textbf{(10.23)}$$

Here, E_{corr}, E_{exact}, and ε represent correlation energy, exact energy of the system, and HF energy, respectively. Since, by variational principle $\varepsilon \geq E_{exact}$, hence, E_{corr} is always negative. The correlation can be further classified as:

Dynamic Correlation

Such a type of correlation arises when two electrons come closer (instantaneous repulsion), the single reference (single Slater determinant) reference methods can take care of this correlation.

Static Correlation

When we consider the excited state of an atom or molecule or the dissociation of a molecule, where more than one Slater determinants are equally important.

To take care of the static correlation several methods have been developed within the *ab initio* and non-*ab initio* method, like, configuration interaction (CI) method, Many-Body Perturbation Theory (MBPT), Coupled cluster (CC) method, Density Functional Theory (DFT), *etc.* Among them CI and DFT belong to the variational method, MBPT is a perturbative technique and CC belongs to neither variational nor perturbative technique. Discussion on these topics is beyond scope of the book. Here we just represent the energies of MBPT (second-order) and DFT.

In the Second Order perturbation theory the energy is expressed as (10.24):

$$E_n^{(2)} = \sum_{m \neq n} \frac{|H'_{mn}|^2}{E_n^{(0)} - E_m^{(0)}} + \int \frac{|H'_{mn}|^2}{E_n^{(0)} - E^{(0)}} dE^{(0)} \qquad \textbf{(10.24)}$$

In the DFT theory, the most accepted theory is the Kohn-Sham (KS) theory. The energy according to KS theory is expressed as (10.25):

$$\left[-\frac{1}{2}\nabla^2 + v_{eff}(r)\right]\psi_i = \varepsilon_i\psi_i \quad \textbf{(10.25)}$$

Here, ∇^2 is Laplacian operator Square, $v_{eff}(r)$ is the effective potential, which is a function of density (ρ) and coordinates (r). The ρ and v_{eff} are defined as:

$$\rho(r) = \sum_i \sum_s |\psi_i(r,s)|^2 \quad \textbf{(10.26)}$$

$$v_{eff}(r) = v(r) + \int \frac{\rho(r\prime)}{[r-r\prime]} dr' + \frac{\delta E_{xc}[\rho]}{\delta\rho(r)} \quad \textbf{(10.27)}$$

Here, v(r) is potential, E_{xc} is the exchange-correlation potential. The exact form is still a challenging task. Details of Density Functional theory can be found in [10].

Here, we shall study using various quantum chemical methods (HF, MP2, and DFT) how the minimum energy and optimized bond length change in nitrogen molecule.

Software Used

Gaussian09 Revision D 01 and Gauss View 5.0.9.

Procedure

1) Nitrogen molecules (N_2) with various bond lengths ranging from 0.7 Å to 4 Å were build-up using Gauss View 5.0.9 software package. Note that the N-N bond distance in every input should differ by 0.1 Å, *i.e.*, 0.7 Å, 0.8 Å, 0.9 Å, 1.0 Å, *etc.* up to 4.0 Å.

2) Run for single-point energy calculation for all these inputs using HF, MP2, and DFT methods. In all the cases we have used a 6-31 G basis set. In this context, please use B3LYP functional to take care of exchange-correlation energy in DFT. The key points will be:

HF: #p hf/6-31g geom=connectivity

MP2: #p mp2/6-31g geom=connectivity

DFT: #p b3lyp/6-31g geom=connectivity

3) Plot variation of Energy (a.u) with respect to Bond Length (Å) (Put Bond Length (Å) along X-axis) for all the three methods and note the minimum energy obtained in individual methods.

4) Take the difference between E_{MP2} and E_{HF} and E_{DFT} and E_{HF}. These differences denote the correlation energy at the minima (Try to plot in MS Excel).

Result

Report energies of N_2 molecule at different bond lengths using the various level of theory in Table **10.8**.

Table 10.8. Energies of N_2 molecule at different bond lengths using the various level of theory.

Bond Length (Å)	E_{HF} (a.u.)	E_{MP2} (a.u.)	Difference (E_{MP2}-E_{HF}) (a.u)	E_{DFT} (a.u.)	Difference (E_{DFT}-E_{HF}) (a.u)
0.7					
....	....	...	....	....	....
4.0					

XIII. WHAT IS AROMATICITY? -- A QUANTUM CHEMICAL OBSERVATION

Theory

According to Hückel's rule of aromaticity planar, cyclic, conjugated molecules with (4n+2) number of π electrons possess aromaticity. Here, n stands for any positive integer including zero. However, in this context cyclic conjugated molecules can be classified as anti-aromatic and non-aromatic compounds. Their classifications are tabulated below in Table **10.9**:

Table 10.9. Comparative study between aromatic, antiaromatic, and non-aromatic compounds.

<table>
<tr><th>Criteria</th><th>Aromatic</th><th>Antiaromatic</th><th>Non-aromatic</th></tr>
<tr><td>Cyclic Compound</td><td>YES</td><td>YES</td><td rowspan="3">Either one among three will be absent</td></tr>
<tr><td>Conjugated system of p orbitals ring of molecule perpendicular to the molecule's plane</td><td>YES</td><td>YES</td></tr>
<tr><td>Planarity</td><td>YES</td><td>YES</td></tr>
<tr><td>π-electron rule</td><td>(4n + 2) π</td><td>4 π</td><td>N/A</td></tr>
</table>

Hence, we can distinguish the difference between these systems. To compute aromaticity, we generally study the ring current in the system. The easiest way to compute the aromaticity is to optimize the molecule first and add a ghost atom (Generally Bq is used in Gaussian software) at the centroid of the molecule and run NMR calculation using gauge-origin including atomic orbitals (GIAO) (Detail discussion about GIAO method is beyond scope of this book) and from nuclear independent chemical shielding (NICS) parameter and calculate the isotropy for each atom (sign reversed). Generally, for aromatic systems, the isotropic values are positive (Hence report negative isotropic value) and for anti-aromatic systems, we get the reverse sign.

Method

DFT method was applied.

Software Used

Gaussian09 Revision D 01 and Gauss View 5.0.9.

Procedure

1) Construct the structure of benzene molecule using GaussView 5 software package.

2) Optimize the structure using Density Functional Theory with a 6-31 G basis set. B3LYP functional is used to take care of the exchange-correlation term.

3) Open the optimized structure and replace the C-atoms in benzene with a ghost atom (Bq).

4) Calculate NMR shielding GIAO parameter for isotropy using the following keyword:

#p nmr=giao b3lyp/6-31+g(d,p) geom=connectivity

5) From the output file note the Anisotropy for all the atoms.

6) Repeat Step 1 to Step 5 for cyclobutadiene and compare the isotropy values with benzene.

Results

Report anisotropy values for benzene and cyclobutadiene in Table **10.10** and **10.11**, respectively.

Table 10.10. Anisotropy values for Benzene (Bq substituted the C atoms).

Atom	Isotropy
H	
H	
H	
H	
H	
H	
C	
C	
C	
C	
C	
C	

Table 10.11. Anisotropy values for Cyclobutadiene (Bq substituted the C atoms).

Atom	Isotropy
H	
H	
H	
H	
C	
C	
C	
C	

Comment

Aromaticity calculation is a very sensitive calculation. So one should repeat the calculations using different methods and basis sets.

XIV. POPULATION ANALYSIS IN MOLECULE

Theory

From the Hartree-Fock theory we can get the energy of a molecule. Using the definition of density, we can get the electron density, which represents the probability of finding an electron in various regions of space, which can be observed by contour plot for various planes drawn through the molecule. However, no such definition exists to define the number of electrons associated with a given atom or nucleus of a bond in a molecule. To get an idea of the same population analysis is useful. As we know a spatial molecular orbital can accommodate a maximum of two electrons, thus, the number of electrons (N) can be written as:

$$N = \sum_{\mu}\sum_{\nu} P_{\mu\nu}S_{\mu\nu} = \sum_{\mu}(PS)_{\mu\nu} = trPS$$

The term $(PS)_{\mu\mu}$ can be represented as the number of electrons to be associated with φ_μ and this formalism is known as Mulliken population analysis [7, 14, 16]. Using population analysis, we can get an idea of electronic charge distribution in a molecule and the bonding, antibonding or nonbonding nature of the molecular orbitals for particular pairs of atoms. To find out how the Mulliken population helps us to study the electronic charge distribution, we shall consider the CO and BF_3 molecule couple and study how population analysis can help us to predict the charge distribution.

Method

DFT method was applied.

Software Used

Gaussian09 Revision D 01 and Gauss View 5.0.9.

Procedure

1) The complex of $CO.BF_3$ was constructed as shown in Fig. **(10.19)** using Gauss View software package and the structure was optimized at the DFT level of theory along with 6-31 G basis set and B3LYP functional (to account exchange-correlation) were applied to compute the optimization process. With the optimized structure, we have separately calculated the BSSE correction and population analysis using "pop=full" in the root section of the Gaussian software package. This keyword (pop) will print the population of the electrons in the bond.

2) The output file was opened and search for the term "Gross orbital populations" to get the population at individual AOs.

3) To find the population on individual atoms search with the keyword "Mulliken Charges".

Fig. (10.19). Structure of CO-BF_3 complex.

Results

1. The Gross orbital population prints as follows:

```
           Gross orbital populations:
                                   1
     1 1   C  1S          1.99782
     2        2S          0.82901
     3        2PX         0.85055
     4        2PY         0.42497
     5        2PZ         0.34821
     6        3S          1.04149
     7        3PX         0.06312
     8        3PY         0.14611
     9        3PZ         0.11996
    10 2   O  1S          1.99774
    11        2S          0.97996
    12        2PX         1.08805
    13        2PY         0.90117
    14        2PZ         0.91819
    15        3S          0.85970
    16        3PX         0.26439
    17        3PY         0.58333
    18        3PZ         0.59976
    19 3   B  1S          1.99604
    20        2S          0.63678
    21        2PX         0.68109
    22        2PY         0.67343
    23        2PZ         0.00616
    24        3S          0.49590
    25        3PX         0.09185
    26        3PY         0.15545
    27        3PZ         0.00772
    28 4   H  1S          0.54850
    29        2S          0.52048
    30 5   H  1S          0.55882
    31        2S          0.52770
    32 6   H  1S          0.55882
    33        2S          0.52774
```

2. The Mulliken Charges for all the atoms are as follows:

Mulliken charges:

		1
1	C	0.178763
2	O	-0.192285
3	B	0.255586
4	H	-0.068980
5	H	-0.086522
6	H	-0.086561

Sum of Mulliken charges = 0.00000

Mulliken charges with hydrogens summed into heavy atoms:

		1
1	C	0.178763
2	O	-0.192285
3	B	0.013523

Note that the overall Mulliken charges are zero, but oxygen atoms having a negative sign, indicating more electron population of oxygen atom than C atom. On the other hand, the B atom having a (+)-ve charge, which implies the B atom has electron deficiency.

3. The population can be visualised as follows:

i) Open the checkpoint file (file.chk) in the GaussView.

ii) Go to Results → Charge Distribution and then select Show Number and Show Vector to visualize the same. The output file is shown below (Fig. **10.20**):

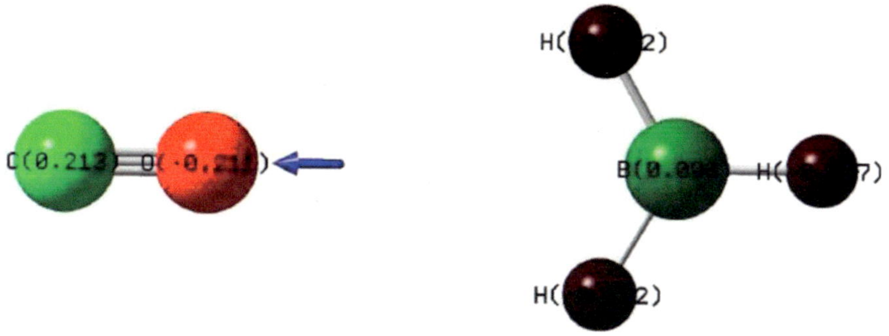

Fig. (10.20). Population analysis on CO-BF3 adduct.

XV. NATURAL BONDING ORBITAL (NBO) ANALYSIS IN A MOLECULE

Theory

In quantum chemistry, the natural bond orbital (NBO) is the calculated orbital having maximum electron density. In this context the NBO is a subset of natural localized orbital. The NBOs are constructed as follows:

Atomic orbital → NAO → NHO → **NBO** → NLMO → MO

Here, NAO, NHO, NLMO are an acronym for natural atomic orbitals, natural hybrid orbitals and natural (semi-)localized molecular orbitals, respectively.

The bonding or anti-bonding NBOs can be represented by the linear combination of directed NHOs and can be represented as:

$$\sigma_{AB/AB*} = C_A h_A \pm C_B h_B$$

AB and AB* represent bonding or antibonding situation, respectively and in the same order + and – signs are also applied. C_is and h_is represent polarization coefficient and NHOs of i atom, respectively. From NBO calculations we get information about a particular orbital, *i.e.,* whether it is bonding, anti-bonding, core state, Rydberg state or Lewis pair, *etc.*

To execute the same, we have tossed propiolamide molecule. This molecule contains a single bond, double bond, triple bond as well as Lewis pair. The geometry of the molecule is shown in Fig. (**10. 21**).

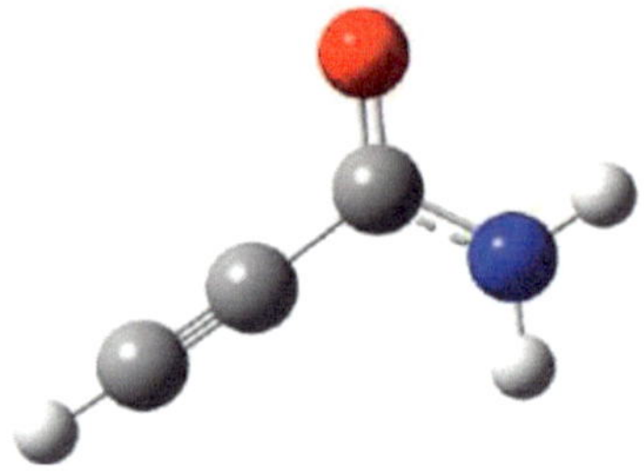

Fig. (10.21). Geometry of propiolamide.

Method

Density Functional Theory (DFT) method was applied.

Software Used

Gaussian09 Revision D 01 and Gauss View 5.0.9.

Procedure

1) The structure of the above-mentioned molecule was built up using the Gauss View software package.

2) The structure was optimized using Density Functional Theory (DFT) method. B3LYP functional was taken into account to take care of the exchange-correlation term. The 6-31G basis set was used as the basis set.

3) The population of electrons was calculated with the optimized geometry using the nbo keyword option and we have saved the NBOs. The overall keyword for the calculation is as follows:

b3lyp/6-31g pop=(nbo,savenbo) geom=connectivity

4) The output file saves the NBO calculation details, they are discussed in the next section.

Result

1) The summary of NBO calculations is summarized below (Only minimum part to explain):

*************************Gaussian NBO Version 3.1****************

NATURAL ATOMIC ORBITAL AND

NATURAL BOND ORBITAL ANALYSIS

***************Gaussian NBO Version 3.1**************************

Natural Bond Orbitals (Summary):

```
                                                          Principal Delocalizations
          NBO                        Occupancy    Energy   (geminal,vicinal,remote)
 ===============================================================================
 Molecular unit  1  (C3H3NO)
   1. BD (   1) C   1 - O   2          1.99943    -1.30783  19(g),47(v),48(v),45(g)
   2. BD (   2) C   1 - O   2          1.99886    -0.50458  50(v),43(g),20(v)
   3. BD (   1) C   1 - N   3          1.99091    -1.11736  48(v),46(g),47(g),49(v)
   4. BD (   1) C   1 - C   6          1.97901    -0.80335  48(g),37(v),51(v),46(v)
                                                    34(v),50(g)
   5. BD (   1) N   3 - H   4          1.99085    -0.90131  45(v),19(v),42(v),21(v)
                                                    27(v),23(v)
   6. BD (   1) N   3 - H   5          1.99171    -0.90085  42(v),50(v)
                                                    44(g),49(r)
   7. BD (   1) C   6 - C   7          1.98013    -1.18082  45(g),51(g),20(v),21(v)
   8. BD (   2) C   6 - C   7          1.98030    -0.43082  42(v),44(v)
                                                    37(g),33(g),19(v),41(v)
   9. BD (   3) C   6 - C   7          1.96591    -0.45674  43(v),22(v)
  10. BD (   1) C   7 - H   8          1.98283    -0.80291  48(g),45(v),33(v)
  11. CR (   1) C   1                  1.99022   -11.21501  48(v),34(v),47(v),46(v)
  12. CR (   1) O   2                  1.99904   -20.10232  19(v),25(g)
                                                    19(g),32(v)
  13. CR (   1) N   3                  1.99012   -15.32102  20(v),21(v),32(v),31(v)
  14. CR (   1) C   6                  1.99856   -11.94085  37(v),51(v),48(g),50(v)
  15. CR (   1) C   7                  1.99826   -11.94139  33(v),45(v),48(g),41(v)
                                                    42(g),19(v),21(v)
  16. LP (   1) O   2                  1.97002    -0.96348  19(v),44(v),45(v),46(r)
                                                    51(g)
  17. LP (   2) O   2                  1.88236    -0.41996  45(v),44(v),20(v),49(r)
  18. LP (   1) N   3                  1.79901    -0.40134  43(v),22(v)
  19. RY*(   1) C   1                  0.01073     1.02275
  20. RY*(   2) C   1                  0.01036     1.00837
  21. RY*(   3) C   1                  0.00191     1.57403
  22. RY*(   4) C   1                  0.00003     1.00103
  23. RY*(   1) O   2                  0.00082     2.07371
  24. RY*(   2) O   2                  0.00028     1.92761
```

```
  24. RY*(   2) O   2                  0.00028     1.92761
  25. RY*(   3) O   2                  0.00011     1.92619
  26. RY*(   4) O   2                  0.00002     3.54601
  27. RY*(   1) N   3                  0.00110     1.91737
  28. RY*(   2) N   3                  0.00079     1.39151
  29. RY*(   3) N   3                  0.00020     1.86598
  30. RY*(   4) N   3                  0.00005     2.25737
  31. RY*(   1) H   4                  0.00120     0.93554
  32. RY*(   1) H   5                  0.00063     0.90169
  33. RY*(   1) C   6                  0.00532     1.44794
  34. RY*(   2) C   6                  0.00220     1.40611
  35. RY*(   3) C   6                  0.00059     1.23137
  36. RY*(   4) C   6                  0.00014     1.04846
  37. RY*(   1) C   7                  0.00542     1.59529
  38. RY*(   2) C   7                  0.00005     1.04951
  39. RY*(   3) C   7                  0.00004     0.98069
  40. RY*(   4) C   7                  0.00001     1.22901
  41. RY*(   1) H   8                  0.00135     0.76119
  42. BD*(   1) C   1 - O   2          0.01123     0.76784
  43. BD*(   2) C   1 - O   2          0.23409     0.14122  50(v),24(g)
  44. BD*(   1) C   1 - N   3          0.05358     0.67053
  45. BD*(   1) C   1 - C   6          0.06689     0.56971
  46. BD*(   1) N   3 - H   4          0.00639     0.69930
  47. BD*(   1) N   3 - H   5          0.00619     0.70308
  48. BD*(   1) C   6 - C   7          0.02307     1.40075
  49. BD*(   2) C   6 - C   7          0.00921     0.28023
  50. BD*(   3) C   6 - C   7          0.01266     0.27445
  51. BD*(   1) C   7 - H   8          0.00810     0.71519
       -------------------------------
              Total Lewis   35.51863  ( 98.6629%)
        Valence non-Lewis    0.43142  (  1.1984%)
        Rydberg non-Lewis    0.04995  (  0.1387%)
       -------------------------------
            Total unit  1   36.00000  (100.0000%)
           Charge unit  1    0.00000
```

The first column (NBO) represents the number of the orbital, the number of appearances, the element symbol, its label in the molecule, the atom to which it is bonded (if applicable), and its label (if applicable). The second column indicates Occupancy (number of electrons) in the orbital. As we know that a bond is formed by 2 electrons, the third column indicates Energy (not in numerical sequence), and the last column indicates Principal Delocalization (geminal, vicinal, remote). The NBO classification of orbitals is tabulated below (Table **10.12**):

Table 10.12. Various terms used in NBO analysis and their meaning.

Designation	Indication	Description
CR	Atomic core State	The core state of the mentioned atom.
BD	Bonding Orbital	The pair of atoms that are associated with this orbital
BD*	Anti-bonding orbital	Same as 2nd row
RY	Rydberg state	A high-lying unoccupied core atomic state
LP	Lewis pair	the associated atom as mentioned

Additionally, bond order can also be calculated with the knowledge of NBO using the formula:

$$\text{Bond order} = \frac{(\text{No. of bonding electrons} + \text{No. of antibonding electrons})}{2}$$

These bond orders correspond to the formal bond orders as shown in Table **10.13**.

Table 10.13. Bond order of bonds.

Bond	Bond Order
C=O	$\frac{(1.99643 + 1.98886 - 0.01123 - 0.23409)}{2} \simeq 2$
C-N	$\frac{(1.99601 - 0.05358)}{2} \simeq 1$

C-C	$\frac{(1.97991 - 0.06689)}{2} \simeq 1$
C-C (Triple bond)	$\frac{(1.98013 + 1.98930 + 1.96591 - 0.02307 - 0.00921 - 0.01266}{2} \simeq 3$

2. Observing the NBOs: Open the checkpoint file (.chk) using Gaussview 5.0 software program suit. Here we mainly focus on the HOMO and LUMO orbitals of the molecule. They are picturized in Fig. **(10.22)**. From our knowledge, we know that the lone pair of electrons from nitrogen forms the HOMO orbital which is coming from the p-orbital of the nitrogen atom. On the other hand, the LUMO is a π* orbital on the C=O bond.

Table 10.14. Entry in output file.

Orbital	Output File				
LUMO	43. BD* (2)C	1-O 2	0.23409	0.14122	50(v), 24(g)
HOMO	18. LP (1)N	3	1.79661	-0.40134	43(v), 22(v)

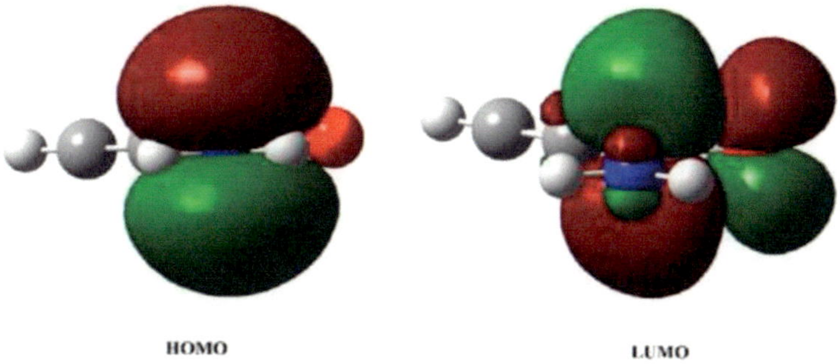

Fig. (10.22). Electron density plot for HOMO and LUMO orbital of propiolamide.

XVI. CALCULATION OF PARAMAGNETIC MOMENT OF A MOLECULE

Theory

In chemistry, the term magnetic moment indicates the magnetic strength and orientation of a magnet or other object that produces a magnetic field. The origin of magnetism arises due to the presence of unpaired electron(s). Electron spin also plays a vital role in the generation of the magnetic field. Based on their relative orientation, we can classify various magnetic materials, like, diamagnetic, paramagnetic, ferromagnetic, antiferromagnetic. For single molecular systems, we generally find that diamagnetic and paramagnetic properties are observed. In diamagnetic materials, the magnetic moment is zero, but for paramagnetic materials, the permanent magnetic dipole moment is observed due to incomplete cancellation of electron spin and orbital magnetic moment. The paramagnetic moment in a molecule can be quantified using quantum chemical methods.

In the gaussian09 software package, we get the Total Magnetic Moment per Atom from the population analysis calculation. In this work, we shall calculate the magnetic moment of the oxygen molecule.

Method

Density Functional Theory (DFT) method was applied.

Software Used

Gaussian09 Revision D 01 and Gauss View 5.0.9.

Procedure

1) The structure of the oxygen molecule was constructed using Gauss View 5 software package.

2) Since the complex can have many possible electronic configurations, so to have the most stable spin structure we have to optimize the structure with individual spin states using 6-31 G (d,p)/B3LYP level of theory (This means that we shall use 6-31 G (d,p) basis set and Density Functional Theory with B3LYP functional).

3) The energies at various spin multiplicities are tabulated and the minimum energy structure is selected for magnetizability calculation.

4) The magnetizability of the most stable structure was obtained using the following command line:

```
# pop=full b3lyp/6-31g(d,p) geom=connectivity
```

Notably, the keyword pop=full allows printing "Gross Orbital Population" in the output file.

5) Open the output file and search for the "Gross Orbital Population"

6) Divide the sum of the "Total" column by the number of atoms.

7) Finally, you obtain the "Total Magnetic Moment per Atom".

Results

Report the variation of energy with respect to spin variation (Table **10.15**)

Table 10.15. Variation of Energy with respect to spin variation.

Spin Multiplicity	Energy (a.u.)	Relative Energy (a.u.)	Relative Energy (kcal/mol)
Singlet		Col 2- Min.Col 2	Col 3 x 627.5
Triplet			
Quintet			
Septet			
Nonet			

The gross orbital population for oxygen molecule in triplet state is shown below:

Gross orbital populations:

Total Alpha Beta Spin

1 1 O1S 1.99314 0.99647 0.99667 -0.00020

2 2S 0.97669 0.50373 0.47296 0.03078

3 2PX 0.96431 0.67402 0.29028 0.38374

4 2PY 0.96431 0.67402 0.29028 0.38374

5		2PZ	0.81118	0.40594	0.40524	0.00070
6		3S	0.94665	0.46266	0.48399	-0.02132
7		3PX	0.52125	0.31913	0.20212	0.11701
8		3PY	0.52125	0.31913	0.20212	0.11701
9		3PZ	0.25563	0.12174	0.13389	-0.01215
10		4XX	-0.00560	-0.00002	-0.00558	0.00556
11		4YY	-0.00560	-0.00002	-0.00558	0.00556
12		4ZZ	0.02791	0.00949	0.01841	-0.00892
13		4XY	0.00000	0.00000	0.00000	0.00000
14		4XZ	0.01445	0.00685	0.00760	-0.00075
15		4YZ	0.01445	0.00685	0.00760	-0.00075
16	2 O	1S	1.99314	0.99647	0.99667	-0.00020
17		2S	0.97669	0.50373	0.47296	0.03078
18		2PX	0.96431	0.67402	0.29028	0.38374
19		2PY	0.96431	0.67402	0.29028	0.38374
20		2PZ	0.81118	0.40594	0.40524	0.00070
21		3S	0.94665	0.46266	0.48399	-0.02132
22		3PX	0.52125	0.31913	0.20212	0.11701
23		3PY	0.52125	0.31913	0.20212	0.11701
24		3PZ	0.25563	0.12174	0.13389	-0.01215
25		4XX	-0.00560	-0.00002	-0.00558	0.00556

26	4YY	-0.00560	-0.00002	-0.00558	0.00556
27	4ZZ	0.02791	0.00949	0.01841	-0.00892
28	4XY	0.00000	0.00000	0.00000	0.00000
29	4XZ	0.01445	0.00685	0.00760	-0.00075
30	4YZ	0.01445	0.00685	0.00760	-0.00075

"Total Magnetic Moment per Atom" can be obtained by the sum of the "Total" column by the number of atoms.

XVII. TO STUDY THE ABSORPTION SPECTRA OF A DYE

Theory

Computation of the excited state of a molecule is an exciting and challenging task for quantum chemists. Because the excitation of molecule involves a multi-reference character (when multiple electronic states having equal weightage), which makes the system a computationally challenging task. A plethora of *ab* initio, as well as first principle methods, has been developed to compute the excited state properties of the molecule. Most of these methods are computationally expensive. However, time-dependent density functional theory (TDDFT) is a comparatively computationally cheaper and accurate method. In this context, although DFT is a variational method so it cannot compute the excited state properties of the molecule. On the other hand, the TDDFT method can calculate the excited state properties and dynamics of molecules. In the TDDFT method, the time-dependent wave function is assumed to be equivalent to the time-dependent electron density. The foundations of the TDDFT method are based on the Runge-Gross (RG) theorem. According to the RG theorem, for an initial wavefunction, there is a unique mapping between the TD external potential of a system as its TD density, which is suggestive of N-body wavefunction (dependent on 3N variable), is equivalent to the density, is dependent upon only 3. Details of the TDDFT method are not discussed here. For details, please see their original work [17].

Further, Time required for an absorption process is 10^{-12} sec which is too fast to reorganize the molecule, which validates the Franck-Condon (FC) principle. According to the FC principle when a molecule is undergoing an electronic transition, such as ionization, the nuclear configuration of the molecule experiences no significant change. This happens due to the fact that the mass of an electron is

~1837 times lighter than the mass of the nucleus. Further, this is a corollary of the Born-Oppenheimer approximation.

Simulation of Absorption Spectra

The UV visible spectra are computed by using a plot of ε *Vs*. λ. Here, ε represents the molar absorption coefficient having unit $M^{-1}cm^{-1}$, and the excitation energies are expressed in wavelength (nm). The observed peak follows a Gaussian band shape having the following relationship:

$$\varepsilon_i(\lambda) = \varepsilon_i^{max} \exp\left[-\left(\frac{\lambda_i-\lambda}{\sigma}\right)^2\right] \quad \textbf{(10.28)}$$

Here, ε_i indicates the molar absorption coefficient of the electronic excitation of interest, λ_i represents ε_i^{max} the wavelength of excitation energy corresponding to the electronic excitation of interest and σ denotes standard deviation in wavenumbers (cm^{-1}), which is related to the width of the simulated band. Precisely, it is the half-width of the Gaussian band.

In this practical, we shall study the excitation/absorption of *o*-, *m*-, and *p*-dinitrobenzene. Further, we shall study the contribution of involved orbitals in an individual excitation in a particular excitation process. The same can be computed using the following formula:

$$Y\% = \frac{x_i^2}{\sum_{i=1}^{n} x_i^2} 100 \quad \textbf{(10.29)}$$

Here, Y% is the single-particle contribution to the vertical excited states and x_i is the single-particle transitions corresponding to a given vertical excited state.

The molar absorption coefficient (ε) can be obtained from the plot as follows:

i) Open the UV-Vis plot using Gauss View software (Results → UV/VIS).

ii) Right-click on the plot and save data with a suitable range and interval of values.

iii) Open the file and obtain the ε value corresponding to λ_{max}.

Software Used

Gaussian09 Revision D 01 and Gauss View 5.0.9.

Procedure

1. The structures of *o*-, *m*-, and *p*- dinitro benzene were drawn using Gauss View 5.0.9 software package.

2. The ground state geometries of all three structures were optimized using Density Functional Theory (DFT) method using the PBE0/6-31 G level of theory. Energies and the lowest three frequency values are noted.

3. With these ground state optimized geometries study the absorption spectra of these molecules using the TD-DFT method. Herein, we have employed CAM-B3LYP functional to take care of the exchange-correlation. We have calculated excitations of ten states. The following keywords are used:

td=(nstates=10) cam-b3lyp/6-31g geom=connectivity

4. The excitation properties (Absorption maxima (λ_{max}), Oscillator Strength (f), Excitation energy, Molar absorption coefficient (ε), and contribution of individual excitation) related to all three molecules are noted.

5. The absorption plot for all three molecules is presented and the λ_{max} values are labeled.

Results

Report ground state energy and frequencies (Table **10.16**) and excitation properties of *o*, *m*, and *p*- dinitrobenzene (Table **10.17**).

Table 10.16. Ground State energy and Frequencies.

Ground State Properties	*o*-dinitro Benzene	*m*- dinitro Benzene	*p*- dinitro benzene
Energy (a.u.)			
Lowest Three Frequencies			

Table 10.17. Excitation Properties of molecules.

Excitation Properties	*o*-dinitro Benzene	*m*- dinitro Benzene	*p*- dinitro Benzene
λ_{max} (nm)			
Oscillator Strength (f)			
Excitation Energy (eV)			
ε (arbitrary unit)			
Orbital contribution (%)			

XVIII. TO STUDY THE ABSORPTION SPECTRA OF A DYE IN SOLUTION

Theory

Computation of excited state of a molecule draws special attraction task for quantum chemists. Because the excitation of molecule involves a multi-reference character (when multiple electronic states having equal weightage), which makes the system a computationally challenging task. A plethora of *ab initio* as well first principle methods have been developed to compute the excited state properties of the molecule. Most of these methods are computationally expensive. However, time-dependent density functional theory (TDDFT) is a comparatively computationally cheaper and accurate method, in this context, although DFT is a variational method so it cannot compute the excited state properties of the molecule but TDDFT method can calculate excited state properties and dynamics of molecules. In the TDDFT method, the time-dependent wave function is assumed to be equivalent to the time-dependent electron density. The foundations of the TDDFT method are based on the Runge-Gross (RG) theorem. According to the RG theorem, for an initial wavefunction, there is a unique mapping between the TD external potential of a system as its TD density, which is suggestive of N-body wavefunction (dependent on 3N variable), is equivalent to the density, is dependent upon only 3. Details of the TDDFT method are not discussed here. For further details, please see their original work [17].

Further, Time required for an absorption process is 10^{-12} sec which is too fast to reorganize the molecule, which validates the Franck-Condon (FC) principle.

According to the FC principle when a molecule is undergoing an electronic transition, such as ionization, the nuclear configuration of the molecule experiences no significant change. This happens because the mass of an electron is almost 1836 times lighter than the mass of the nucleus.

Solvent plays a vital role in influencing the absorption spectra of molecules. Depending on the type of solvent (polar or non-polar) and mode of excitation from HOMO to LUMO and nature of solute, the changes in absorption and emission spectra take place. It is commonly observed that the same solute exhibits two different λ_{max} in two different solvents. Thus, the choice of a suitable solvent is important. Generally, such solvents are transparent in the UV-Visible region (does not absorb in the UV-Vis range), and less polar (minimise the interaction with the solute molecule). Examples of such solvents include water, methanol, ethanol, *etc.* In presence of a solvent, the electronic transition shows the following order:

$$n\rightarrow\pi^* < \pi \rightarrow \pi^* < n \rightarrow \sigma^* < \sigma \rightarrow \pi^* < \sigma \rightarrow \sigma^* \qquad [18]$$

In quantum chemistry, the entire solvent model assumes that the solute is trapped in the cavity within the solvent reaction field. Several solvent models are included in the G09 software package, like Polarizable Continuum Model (PCM), Integral Equation Formalism Variant (IEFPCM), *etc.* In the IEFPCM model, the solute cavity is formed by overlapping various spheres.

In this practical, we shall study the excitation properties of *o-, m-, and p-* dinitrobenzenes in solvent phase. Further, we shall study the contribution of involved orbitals in an individual excitation in an excitation process using Equation (10.29). To find the effect of solvent we have used ethanol as solvent and study the effect of solvent on the excitation properties of molecules.

Please note that the effect of solvents is widely used to study molecules in the ground state as well as the emission properties of molecules.

Software Used

Gaussian09 Revision D 01 and Gauss View 5.0.9.

Procedure

1. The structures of *o*-, *m*-, and *p*- dinitro benzene were drawn using Gauss View 5.0.9 software package.

2. The ground state geometries of all three structures were optimized using Density Functional Theory (DFT) method using the PBE0/6-31G level of theory. Energies and the lowest three frequency values are noted.

3. With these optimized geometries study the absorption spectra of these molecules using the TD-DFT method. Herein, we have employed CAM-B3LYP functional to take care of the exchange-correlation. We have calculated excitations of ten states. The calculations were performed in both the gaseous phase as well as in the solvent phase. For solvent phase excitation energy calculations, the following keywords are used:

#td = (nstates=10) cam-b3lyp/6-31g scrf=(CPCM,solvent = ethanol) geom = connectivity

4. The excitation properties (Absorption maxima (λ_{max}), Oscillator Strength (f), Excitation energy, Molar absorption coefficient (ε), and contribution of individual excitation) related to all three molecules are noted for both gas phase and solvent phase.

5. The absorption plots for all three molecules are presented and the λ_{max} values are labeled.

Results

Report ground state energy and frequencies (Table **10.18**) and excitation properties of the molecules (Table **10.19**).

Table 10.18. Ground State energy and Frequencies.

Ground State Properties	***p*- dinitro Benzene**
Energy (a.u.)	
Lowest Three Frequencies	

Table 10.19. Excitation Properties of molecules.

Excitation Properties	***p*- dinitro benzene**	
	Gas	**Solvent**
λ_{max} (nm)		
Oscillator Strength (f)		
Excitation Energy (eV)		
ε (arbitrary unit)		
Orbitals involved in λ_{max}		
Orbital contribution (%)		

XIX. TO STUDY THE EMISSION SPECTRA OF A DYE

Theory

Computation of the excited state of a molecule is an exciting and challenging task for quantum chemists. Because the excitation of molecule involves a multi-reference character (when multiple electronic states having equal weightage), which makes the system a computationally challenging task. A plethora of *ab initio* as well first principle methods have been developed to compute the excited state properties of the molecule. Most of these methods are computationally expensive. However, time-dependent density functional theory (TDDFT) is a comparatively computationally cheaper and accurate method, In this context, although DFT is a variational method so it cannot compute the excited state properties of the molecule TDDFT can compute the excited state properties of the molecule.

Notably using Franck Codon principle, the geometry of the molecule does not change during the adsorption process because the time required for electronic absorption is much faster than the nuclear movement, but when the molecule emits light (fluorescence/phosphorescence) in this case potential energy surface of the excited state shift from the ground state and hence the emission maxima is higher than the same for absorption maxima. This can be expressed as:

$$(\lambda max)_{emission} > (\lambda max)_{absorption}$$

However, the pattern for absorption and emission will be the same which can be explained by the mirror image rule. In this context some lifetime range related to electronic excitation is reported in Table **10.20** below (These Data are taken from Ref [18]):

Table 10.20. Life-time of various de-excitation processes.

Process	Identity	Time (Sec)
Excitation	Absorption	10^{-15} Sec
Radiative Emission	Fluorescence	10^{-10} -10^{-7} Sec
	Phosphorescence	10^{-6} -1 Sec
Non-Radiative Emission	Vibrational Relaxation	10^{-12} -10^{-10} Sec
	Intersystem Crossing	10^{-10} -10^{-8} Sec
	Internal Conversion	10^{-11} -10^{-9} Sec

In this work, we shall compute the emission spectra of *p*-dinitrobenzene using TDDFT. Further, we shall study the contribution of involved orbitals in an individual excitation in a particular excitation process. The same can be computed using the following formula:

$$Y\% = \frac{x_i^2}{\sum_{i=1}^{n} x_i^2} 100 \qquad \textbf{(10.30)}$$

Here, Y% is the single-particle contribution to the vertical excited states and x_i is the single-particle transitions corresponding to a given vertical excited state.

The molar absorption coefficient (ε) can be obtained from the plot as follows:

i) Open the UV-Vis plot using Gauss View software (Results → UV/VIS).

ii) Right-click on the plot and save data with a suitable range and interval of values.

iii) Open the file and obtain the ε value corresponding to λ_{max}.

Software Used

Gaussian09 Revision D 01 and Gauss View 5.0.9.

Procedure

1) The molecule is drawn using Gauss view software 5.0.9.

2) Optimize the structure of the ground state using the B3LYP/6-31 G level of theory. Ensure the accomplishment of true minima by the absence of negative frequency. This geometry corresponds to ground state optimized geometry.

3) Open the optimized geometry and do the further calculation as follows: select optimization to a minimum (actually it is recommended to calculate of frequency in the excited state using TDDFT method is not implanted in Gaussian 09, but successfully implemented in Gaussian 16 software package). Hence, we have opted for the only optimization of the excited state.

4) In the method section, we have chosen TD-SCF in combination with the DFT method.

5) CAM-B3LYP functions are chosen as an exchange-correlation functional with generalized gradient approximation (GGA). The CAM-B3LYP function is chosen because this function is well established functional to compute the excited state properties of molecules. After all, it takes care of long-range and short-range interaction properly.

6) 6-31G+(d,p) basis set was chosen. Only six excited states have been solved using the TDDFT method. The input file keywords should be as follows:

opt td=(nstates = 10) cam-b3lyp/6-31g geom = connectivity

7) Compare the change in geometry from the ground state and excited state by comparing the geometry parameters (bond length, bond angle, and dihedral angles).

8) Open the output file using Gauss view software 5.0.9, the spectral data corresponds to the first iteration indicates the absorption spectra. The computed spectral data corresponds to the last iteration represents the emission spectra of the given molecule. Note the λ_{max}, Oscillator strength (f), excitation energy (ΔE) for both absorption and emission process.

9) Plot the simulated absorption and emission spectra of the dye.

Results

Report the results in Table **10.21** and **10.22** and plot the simulated absorption and emission spectra.

Table 10.21. Difference in geometry from the ground state to excited state.

Significant Parameter	Ground State	Excited State	Difference
Bond Length (Å)			
Bond Angle (°)			
Dihedral Angle (°)			

Table 10.22. Excited state properties in absorption and emission process.

Properties	Absorption	Emission
Wavelength Maxima (λ_{max}) (nm)		
Oscillator Strength (f)		
Excitation Energy (ΔE) (eV)		
Molar Absorption Coefficient (ε)		
Percentage of the contribution of orbital		

XX. TO FIND OUT THE TRANSITION STATE OF AN S_N2 AND ALSO TO FIND OUT THE REACTION PATH BY INTRINSIC REACTION COORDINATE

Theory

To find out the reaction mechanism in chemistry is a very important as well as very difficult task. All the theories have some drawbacks. Among them, collision theory and transition state theory are most accepted. Collision theory is based on classical law in coordination with the statistical thermodynamic method. However, this method is suffering from several flaws, like it does not take into account the fact that proper molecule orientation is the requirement for a chemical reaction to occur. Active collision does not occur due to lack of proper orientation and therefore the experimentally observed frequency is less than determined based on the theory of collision, *etc.* On the other hand, Transition State Theory (TST) is based on a purely quantum mechanical method and this is the most widely accepted method. It was developed by Polanyi. Although, this method cannot successfully explain several phenomena like it cannot justify how the yield of product changes with the change of concentration of reactant. To circumvent the problems, several methods have been developed within the TST framework. In TST, a non-linear molecule has (3N-6) vibrational coordinates. When two reactants form one product then we have 2*(3N-6) coordinates on the reactant side and (3N-6) on the product side. Additionally, each of the molecules has a 3 coordinate system that means this reaction becomes complicated. So, we assume the reaction proceeds *via* some particular coordinate axis and freeze that coordinate, so our calculation becomes easier. According to TST, the reaction process is initiated by Van der Waal's force of attraction between the reactant in their P.E surface. Near the transition state, the bond-breaking process started, and just after the transition state, bond formation takes place and the transition state lies in between which is indicated as a maxima point containing a single imaginary frequency. The whole reaction path can be described by Intrinsic Reaction Coordinate (IRC), as developed by Fukui [19] which adopts a local optimisation process, following a steepest closest method.

In this practical, we shall search the transition state of the following reaction.

$$CH_3Cl + Br^- \longrightarrow CH_3Br + Cl^-$$

To search the transition state, we adopt Berny's T.S search process and the reactant and product can be fetched by the IRC method.

Procedure

1) The reactants are constructed using Gauss View 5.0.9 software package. Since the structure of the transition state is not the same as the optimised structure, so we should guess the structure of the molecule near the transition state in such a way that it will be a good guess to start the transition state calculation.

2) To find out the transition state we have used the following keyword:

opt=(calcall,ts) freq b3lyp/6-31g geom=connectivity nosymm

Note that here nosymm keyword is used to indicate symmetry option is dropped. The keyword calcfc will allow the computer to calculate force constant in each iteration. This is very necessary for TS calculation. To avoid computational cost one may use calcfc instead of calcall, this will calculate force constant at the final step, instead of in every iteration.

3) The transition state is confirmed by the presence of a single imaginary frequency in the output and also checking the displacement vector. Note the energy of the transition state. To ensure a true transition state is achieved, we have to f*etc*.h the reactant and product *via* Intrinsic Reaction Coordinate. For this purpose, copy the transition checkpoint file (file.chk) as TS-for.chk and TS –rev.chk. Since the reaction can move both in forward and reverse direction. So, IRC will try to fetch both pathways. Hence, we have to make two different files for IRC. One indicates the reaction towards the reactant side (IRC-Reverse) and another is indicative of IRC towards the product side (IRC-Forward) and accordingly, input files for IRC should be arranged. The input file for IRC-Forward is:

% chk = Ts- for.chk

#p irc=(calcall,tight,maxcycle=50,stepsize=-3,maxpoints=999,forward) b3lyp/6-31g(d,p) guess(read)

The input file for IRC-Reverse is:

% chk = Ts- rev.chk

#p irc=(calcall,tight,maxcycle=50,stepsize=-3,maxpoints=999,reverse) b3lyp/6-31g(d,p) guess(read)

Individual calculations were performed to take care of the forward and reverse path of the reaction to achieve reactant and product.

Note

To do so, we do IRC calculation for forward and backward independently but it is not guaranteed that we will have desired product.

4) To obtain the thermochemistry, reaction barrier of the reaction, we have to optimise the reactant and product separately to a true minimum and find the thermochemistry, reaction barrier.

Result

Report the energy, Gibb's free energy, and enthalpy (Table **10.23**).

Information Related to the Transition State

The energy of TS:

Negative Frequency value of TS.

Save the image of the normal mode of vibration that corresponds to the negative frequency along with displacement vectors (Fig. **10.23**).

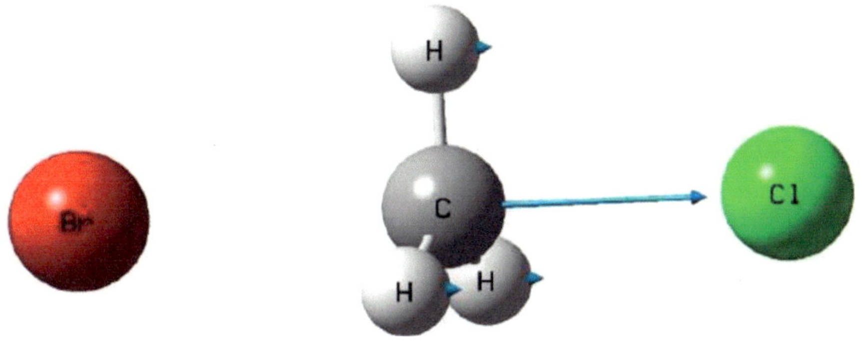

Fig. (10.23). Normal mode of vibration that corresponds to the negative frequency along with displacement vectors (showed by arrow).

Table 10.23. Energy, Gibbs' Free Energy, Enthalpy of reaction (Zero-point Energy (ZPE) corrected Gibbs' Free Energy and Enthalpy is chosen).

Parameter	Reactant	TS	Product	Factors
Energy (a.u.)	-	-	-	Barrier height= E_{TS}-$E_{Reactant}$
Gibbs' Free Energy	-	-	-	$\Delta G = G_{Product} - G_{Reactant}$
Enthalpy	-	-	-	$\Delta H = H_{Product} - H_{Reactant}$

Note: Here Gibbs' Free energy and Enthalpy indicate Sum of electronic and thermal Free energy and Sum of electronic and thermal entropy.

XXI. TO FIND OUT STRUCTURAL DETAILS OF A CRYSTAL FROM CRYSTALLOGRAPHIC INFORMATION FILE (CIF)

Theory

In a crystal, molecule/atoms are oriented in a very specific arrangement with a certain boundary condition which is known as a lattice parameter. For a specific crystal, these lattice parameters are fixed. For theoretical or computational studies, this is the first information to start with. Experimentally, using X-ray diffraction (XRD) methods are used to study the structure of crystals. Based on the type of crystal, various XRD methods have been adopted, like single-crystal XRD, powdered XRD, *etc.* These output data are very useful for a computational chemist as input to study a particular crystal. The crystal data are generally obtained in crystallographic information file (The files are generally named as filename.cif). In this context, a huge number of cif data are available in Crystallography Open Database [20] for absolutely free. In a crystallographic information file, we can get the following important parameters:

(A) Lattice parameters of the crystal. This includes the length of the crystal along the X, Y, and Z coordinate axis and the angle created between them.

(B) Space group of the crystal which gives information about the symmetry properties in the crystal.

(C) Arrangement of the atoms in a crystal expressed in terms of fractional coordinates. Notably, the fractional coordinate is another way of representing the coordinates of atoms, but the difference is in cartesian coordinate. The molecules are represented as they are located in the X, Y, Z axis, but when we represent the coordinates of atoms in a crystal in terms of fractional coordinates, then we consider the origin of the crystal is situated at (0,0,0) and edge length of each side is unity and the atoms are packed inside the crystal. Hence, all the atoms will have a fractional coordinate.

Software Required

Gauss View 5.0.9 software package or VESTA [21] (This is a free visualization software).

Procedure

1) The provided cif of a particular crystal is opened in Gauss view 5.0.9 or VESTA software package.

2) In Gauss View go to the edit section and click on PBC to open the Periodic Boundary Conditions (PBC) of the particular crystal and by enabling the Space group symmetry, we can get information on the space group of the lattice. Information about the cell can be obtained by clicking the Cell tab. Here, we can get information about cell dimensions, *i.e.*, a, b and c as well as the interaxial angles, *i.e.*, α, β and γ. The cell lengths are given in Å and the bond angles are given in Degree (°). The periodicity of the system can be observed by clicking the 'view' section and the system can be expanded to any direction along with a, b or c. For a better view, please select Replicate Content Display to Higher Layer. To cut the crystal to a suitable (hkl) plane, select on the reduce section and cut the cell according to your choice. Note down the space group and cell parameters of the given crystal.

3) To get the coordinates of the crystal using Gauss Viewgo to Results and select View File. The fractional coordinates of elements will appear. Publication details are also available here.

4) To obtain the same information about the crystal using VESTA, open the cif in the VESTA visualization software package. Go to Edit → Edit data → Unit Cell and get the information of the type of crystal and the space group of the Crystal as well as the lattice parameters, *i.e.*, a, b, c and α, β, γ. Like Gauss View here also the cell lengths are given in Å and the bond angles are given in Degree (°). Details about the fractional coordinate position of atoms are available in the 'Structure parameters' section. The crystal can be reduced to some (hkl) plane by clicking on the crystal shape. The lattice spacing (d) can be obtained from here. To validate the periodicity of the crystal, select Objects followed by Volumetric data and expand somewhat value to any direction along the coordinate axis. Note down the Space group andCell parameters of the given crystal using VESTA.

5) To get the publication details, open the file in text editor files. The date is omitted sometimes.

Results

Take the crystal structure of NaCl. **COD ID**: 1000041. Open the file in g09. Note down the following:

Publication details:

Space group =

Report the cell parameters (Table **10.24**) and fractional coordinates (Table **10.25**) of the crystal.

Table 10.24. Cell parameters of the given crystal.

a (Å)	b (Å)	c (Å)	d (Å)	α (⁰)	β (⁰)	γ (⁰)

Table 10.25. Fractional Coordinate of the elements in the crystals.

Element	x	y	Z
Element 1 (Na)			
Element 2 (Cl)			

REFERENCES

[1] PerkinElmer (http://www.cambridgesoft.com/pki/)

[2] Dennington, R.; Keith, T.; Millam, J. *Semichem Inc.*, Shawnee Mission, KS, Gauss View, Version 5, **2009**.

[3] Schaftenaar, G.; Noordik, J.H., *J. Comput. -Aided Mol. Design*, **2000,** 14, 123.

[4] Frisch, M. J.; Trucks, G. W.; Schlegel,H. B.; Scuseria, G. E.; Robb, M. A.; Cheeseman, J. R.;Scalmani,G.; Barone,V.;Mennucci,B.;Petersson,G. A.;Nakatsuji,H.;Caricato,M.; Li, X.; Hratchian, H. P.; Izmaylov, A. F.; Bloino, J.; Zheng, G.; Sonnenberg, J. L.;Hada, M.;Ehara, M.; Toyota,K.; Fukuda, R.; Hasegawa, J.; Ishida, M.; Nakajima, T.; Honda, Y.; Kitao, O.; Nakai,H.; Vreven, T.; Montgomery, J. A. Jr.; Peralta, J. E.; Ogliaro, F.; Bearpark, M.; Heyd, J. J.; Brothers, E.; Kudin, K. N.; Staroverov, V. N.; Kobayashi, R.; Normand, J.; Raghavachari, K.; Rendell, A.; Burant, J. C.; Iyengar, S. S.; Tomasi, J.; Cossi, M.; Rega, N; Millam, N. J.; Klene, M.; Knox, J. E.; Cross, J. B.; Bakken, V.; Adamo, C.; Jaramillo, J.; Gomperts, R.; Stratmann, R. E.; Yazyev, O.; Austin, A. J.; Cammi, R.; Pomelli, C.; Ochterski, J. W.; Martin, R. L.; Morokuma, K.; Zakrzewski, V. G.; Voth, G. A.; Salvador, P.; Dannenberg, J. J.; Dapprich, S.; Daniels, A. D.; Farkas, Ö.; Foresman, J. B.; Ortiz, J. V.; Cioslowski, J.; Fox D. J. Gaussian, Inc., Wallingford CT, **2009**. Gaussian 09.

[5] TURBOMOLE V6.2 2010, a development of University of Karlsruhe and Forschungszentrum Karlsruhe GmbH, 1989-2007, TURBOMOLE GmbH, since **2007**; available from http://www.turbomole.com.

[6] Schmidt, M. W.; Baldridge, K. K.; Boatz, J. A. Elbert, S. T. Gordon, M. S. Jensen, J. H. Koseki, S. Matsunaga, N. Nguyen, K. A. Su, S. J. Windus, T. L. Dupuis M., Montgomery, J. A. *J. Comput. Chem.***1993**14, 1347.
[7] (a) Neese, F. *Wiley Interdiscip. Rev.: Comput. Mol. Sci.*, **2012,**2, 73. (b) Neese, F. *Wiley Interdiscip. Rev.: Comput. Mol. Sci.*, **2017,** 8, e1327.
[8] Szabo, A.; Ostlund, N. S. '*Modern Quantum Chemistry'*, McGraw-Hill, New York, **1989**.
[9] (A) Cizek. J. *Adv. Quant. Chem.***1969,**14, 35. (B) Bartlett, R. J. '*Modern Electronic Structure Theory, Part II, Advanced Series in Physical Chemistry*' Vol. 2, edited by Yarkony, D. R., World Scientific, Singapore, **1995**, p1047. (C) Bartlett R. J., *Annu. Rev. Phys. Chem.* **1981** 32, 359. (D) Shavitt, I; Bartlett, R. J. "*Many-Body Methods in Chemistry and Physics*" Cambridge University Press, **2009**.
[10] (A) Parr, R. G.; Yang, W. '*Density functional theory of atoms and molecules*', (Oxford University press, New York, **1989**. (B) Koch, W. Holthausen, M. C. "*A Chemist's Guide to Density Functional Theory*" Willey Online Library, **2001**. (C) Bamzai, A. S.; Deb, B. M. *Rev. Mod. Phys.* **1981,** 53, 95.
[11] Born, M.; Oppenheimer, R. J., *Ann. Physik.***1927**, 84, 457
[12] Levine, I. N. "*Quantum Chemistry*" Prentice Hall of India, New Delhi,
[13] McQuarrie, D. A., Simon, J. D. "*Physical Chemistry A Molecular Approach*", University Science Books Sausalito, California. **1997**.
[14] Harris, J. *Phys. Rev.* B, 1985, 31, 1770.
[15] Boys S. F.; Bernardi, F. *Mol. Phys.*, **1970**, 19, 553.
[16] Mulliken, R.S. *J. Chem. Phys.* **1955,** 23, 1833.
[17] Runge, E.; Gross, E. K. U. *Phys. Rev. Lett.***1984,** 52, 997.
[18] Valeur, B. "*Molecular Fluorescence: Principles and Applications*" Wiley-VCH Verlag GmbH, ISBNs: 3-527-29919-X (Hardcover); 3-527-60024-8 (Electronic).
[19] Fukui, K. *Acc. Chem. Res.*, **1981**, 14, 363.
[20] Crystallography Open Database (http://www.crystallography.net/cod/).
[21] JP Minerals (https://jp-minerals.org/vesta/en/download.html)

SUBJECT INDEX

A

B

Charu Arora and Sumantra Bhattacharya

F

L

M

N

O

P

Q

R

S

T

U

W

X

Made in the USA
Las Vegas, NV
03 November 2026

57421906R00209